Children and Young People as Knowledge Producers

Despite the widespread promotion of children's voices by activists and policy-makers over the last decade, the potential for young people's knowledge to impact on adult agendas and policy arenas is by no means a certainty. This book presents critiques of participation in settings where young people are the centre of attention. The complexities and power-dynamics of youth-adult relationships are observed and analysed in a wide diversity of study environments, from Hull to Sao Paulo, rural Lesotho to Ghana, using varied methods and over different time frames, but with a strong focus throughout on context, practice, impacts and associated ethical considerations. The central concern of the book is not whether young people can produce better knowledge than adults, but rather how to better understand the different types of knowledge which emerge from diverse actors within different generations, in order to ensure that the maximum benefits accrue to children and young people, with and for whom, the research is conducted.

This book was originally published as a special issue of *Children's Geographies*.

Gina Porter is a Senior Research Fellow in the Department of Anthropology at Durham University, UK. She has been conducting research into participatory methodologies for many years, principally in sub-Saharan Africa. Her recent work has focused on the co-production of knowledge with children and young people in Ghana, Malawi and South Africa, and with older people in Tanzania. She is currently leading an interdisciplinary research study on the impact of mobile phones on young people's lives in Ghana, Malawi and South Africa.

Janet Townsend is a Visiting Fellow at the School of Geography, Politics and Sociology, Newcastle University, UK. She is a feminist who has engaged in participatory research with poor women in low income countries. She is concerned with issues of poverty, power, self-empowerment and the (dangerous) power of academics, particularly those in prosperous countries.

Kate Hampshire is a Reader in the Department of Anthropology, a Lecturer in Health and Human Sciences and a Fellow of the Wolfson Research Institute for Health and Wellbeing at Durham University, UK. She works on children and young people's health and wellbeing in various settings, using participatory research approaches. Her recent research includes (with Gina Porter) co-production of knowledge with young people in Ghana, Malawi and South Africa, children's use of medicines in Ghana, and social wellbeing among school children in Northeast England.

Children and Young People as Knowledge Producers

Edited by
Gina Porter, Janet Townsend and
Kate Hampshire

LONDON AND NEW YORK

First published 2014
by Routledge

2 Park Square, Milton Park, Abingdon, Oxfordshire OX14 4RN
711 Third Avenue, New York, NY 10017

Routledge is an imprint of the Taylor & Francis Group, an informa business

First issued in paperback 2018

Copyright © 2014 Taylor & Francis

All rights reserved. No part of this book may be reprinted or reproduced or utilised in any form or by any electronic, mechanical, or other means, now known or hereafter invented, including photocopying and recording, or in any information storage or retrieval system, without permission in writing from the publishers.

Notice:
Product or corporate names may be trademarks or registered trademarks, and are used only for identification and explanation without intent to infringe.

British Library Cataloguing in Publication Data
A catalogue record for this book is available from the British Library

ISBN 13: 978-0-415-74065-4 (hbk)
ISBN 13: 978-1-138-38319-7 (pbk)

Typeset in Times New Roman
by Taylor & Francis Books

Publisher's Note
The publisher accepts responsibility for any inconsistencies that may have arisen during the conversion of this book from journal articles to book chapters, namely the possible inclusion of journal terminology.

Disclaimer
Every effort has been made to contact copyright holders for their permission to reprint material in this book. The publishers would be grateful to hear from any copyright holder who is not here acknowledged and will undertake to rectify any errors or omissions in future editions of this book.

Table of Contents

Citation Information — vii
Notes on Contributors — ix

Introduction: Children and young people as producers of knowledge
Gina Porter, Janet Townsend and Kate Hampshire — 1

1. 'It came up to here': learning from children's flood narratives
   Marion Walker, Rebecca Whittle, Will Medd, Kate Burningham, Jo Moran-Ellis and Sue Tapsell — 5

2. Emerging relationships and diverse motivations and benefits in participatory video with young people
   Matej Blazek and Petra Hraňová — 21

3. Learning from young people about their lives: using participatory methods to research the impacts of AIDS in southern Africa
   Nicola Ansell, Elsbeth Robson, Flora Hajdu and Lorraine van Blerk — 39

4. Critical dialogue, critical methodology: bridging the research gap to young people's participation in evaluating children's services
   Liz Todd — 57

5. What we say and what we do: reflexivity, emotions and power in children and young people's participation
   Victoria Jupp Kina — 71

6. Taking the long view: temporal considerations in the ethics of children's research activity and knowledge production
   Kate Hampshire, Gina Porter, Samuel Owusu, Simon Mariwah, Albert Abane, Elsbeth Robson, Alister Munthali, Mac Mashiri, Goodhope Maponya and Michael Bourdillon — 89

Index — 103

Citation Information

The chapters in this book were originally published in *Children's Geographies*, volume 10, issue 2 (May 2012). When citing this material, please use the original page numbering for each article, as follows:

Introduction
Editorial: Children and young people as producers of knowledge
Gina Porter, Janet Townsend and Kate Hampshire
Children's Geographies, volume 10, issue 2 (May 2012)
pp. 131–134

Chapter 1
'It came up to here': learning from children's flood narratives
Marion Walker, Rebecca Whittle, Will Medd, Kate Burningham,
Jo Moran-Ellis and Sue Tapsell
Children's Geographies, volume 10, issue 2 (May 2012)
pp. 135–150

Chapter 2
Emerging relationships and diverse motivations and benefits in participatory video with young people
Matej Blazek and Petra Hraňová
Children's Geographies, volume 10, issue 2 (May 2012)
pp. 151–168

Chapter 3
Learning from young people about their lives: using participatory methods to research the impacts of AIDS in southern Africa
Nicola Ansell, Elsbeth Robson, Flora Hajdu and
Lorraine van Blerk
Children's Geographies, volume 10, issue 2 (May 2012)
pp. 169–186

CITATION INFORMATION

Chapter 4
Critical dialogue, critical methodology: bridging the research gap to young people's participation in evaluating children's services
Liz Todd
Children's Geographies, volume 10, issue 2 (May 2012)
pp. 187–200

Chapter 5
What we say and what we do: reflexivity, emotions and power in children and young people's participation
Victoria Jupp Kina
Children's Geographies, volume 10, issue 2 (May 2012)
pp. 201–218

Chapter 6
Taking the long view: temporal considerations in the ethics of children's research activity and knowledge production
Kate Hampshire, Gina Porter, Samuel Owusu, Simon Mariwah, Albert Abane, Elsbeth Robson, Alister Munthali, Mac Mashiri, Goodhope Maponya and Michael Bourdillon
Children's Geographies, volume 10, issue 2 (May 2012)
pp. 219–232

Please direct any queries you may have about the citations to clsuk.permissions@cengage.com

Notes on Contributors

Albert Abane, Department of Geography and Regional Planning, University of Cape Coast, Cape Coast, Ghana

Nicola Ansell, Centre for Human Geography, Brunel University, London, UK

Matej Blazek, School of the Environment, University of Dundee, Dundee, UK

Lorraine van Blerk, Geography, School of Environment, University of Dundee, Dundee, UK

Michael Bourdillon, Department of Sociology, University of Zimbabwe, Harare, Zimbabwe

Kate Burningham, Department of Sociology and Centre for Environmental Strategy, University of Surrey, Guildford, UK

Flora Hajdu, Department of Urban and Rural Development, Swedish University of Agricultural Sciences, Uppsala, Sweden

Kate Hampshire, Department of Anthropology, Durham University, UK

Petra Hraňová, Civic Association Ulita, Bratislava, Slovakia

Victoria Jupp Kina, School of Health Sciences and Social Care, Brunel University, Uxbridge, UK

Goodhope Maponya, CSIR, Pretoria, South Africa

Simon Mariwah, Department of Geography and Regional Planning, University of Cape Coast, Cape Coast, Ghana

Mac Mashiri, Independent Transport Consultant, Pretoria, South Africa

Will Medd, Lancaster Environment Centre, Lancaster University, Lancaster, UK

Jo Moran-Ellis, Department of Sociology, University of Surrey, Guildford, UK

Alister Munthali, Centre for Social Research, Chancellor College, University of Malawi, Zomba, Malawi

Samuel Owusu, Department of Population and Health, University of Cape Coast, Cape Coast, Ghana

Gina Porter, Department of Anthropology, Durham University, UK

NOTES ON CONTRIBUTORS

Elsbeth Robson, Department of Geography, Environment and Earth Sciences, University of Hull, Hull, UK

Sue Tapsell, Flood Hazard Research Centre, Middlesex University, London, UK

Liz Todd, School of Education, Communication and Language Sciences, Newcastle University, Newcastle upon Tyne, UK

Janet Townsend, Department of Geography, Newcastle University, UK

Marion Walker, Lancaster Environment Centre, Lancaster University, Lancaster, UK

Rebecca Whittle, Lancaster Environment Centre, Lancaster University, Lancaster, UK

INTRODUCTION
Children and young people as producers of knowledge

The 1989 UN Convention on the Rights of the Child represented a particularly important way-mark for child-centred studies because it affirmed children's rights to participation: the right to give and receive information, rights of association and rights to participation in cultural life. Since then the potential for young people to participate in a range of other communication and advocacy activities, including a more proactive role in participatory research, has been promoted with growing determination by many child-focused non-governmental organisations (NGOs) and academics. Concepts of children's rights and empowerment are central to these efforts. Save the Children's briefing paper (2000) on research, monitoring and evaluation with children and young people puts the emphasis firmly on partnership with children – the importance of not only collaborative work between children and adults, but also on allowing children to plan and carry out their own research.

However, despite the widespread promotion of children's voices by activists and policy-makers in recent years, the potential for young people's knowledge to impact on adult agendas and policy arenas remains less than certain. For academics, working with children as research partners (as opposed to research subjects) is by no means beyond dispute. An exciting but arguably perilous enterprise, it brings to the fore a range of debates around power relations, ethics, capacities and competencies (of all concerned). James (2007) asks whether research carried out by children necessarily represents a more accurate or authentic account of children's issues: her warning about the dangers of ethnographic ventriloquism may be salutary and is specifically considered in a number of the papers that comprise this special issue.

Most of the papers in this collection were presented, in initial form, in a session on Children and Young People as Knowledge Producers in August 2009, at the Royal Geographical Society/Institute of British Geographers (RGS/IBG) annual conference (which had as its overall theme Geography, Knowledge and Society) at Manchester University, UK. The authors offer diverse disciplinary and interdisciplinary perspectives on young people's role and potential in knowledge production, in a range of contexts: two are set in UK (Todd, Walker *et al.*), one in Slovakia (Blazek and Hraňová), one in Brazil (Jupp-Kina) and two in sub-Saharan Africa (Ansell *et al.*, Hampshire *et al.*). In terms of disciplinary focus, the majority of authors, while based in departments of anthropology, education, geography and psychology, are academics involved in interdisciplinary endeavours in their child- and youth-focused research, whether in UK or overseas. While at least one of the academic researchers has also had some prior experience as a social-work practitioner (Jupp-Kina), the value of academic/practitioner collaboration is clearly demonstrated in the paper by Blazek and Hraňová, which brings together the reflections of an academic geographer and a community youth worker.

Although they employ diverse methods, everyone in this special issue explores 'participatory' work with children and young people, using critical perspectives of children's production of knowledge and seeing children mainly as social actors in their own right. This leads not only to very valuable insights into issues often viewed primarily from an adult perspective – impacts of flooding (Walker *et al.*), evaluation of educational psychology services (Todd) – but, of likely particular interest to the readers of this journal, a focus on critiques of participation in settings where young people are the central focus of attention. As Hampshire *et al.* observe (this collection), there has been a gradual move in social science away from seeing children and young people as mere 'objects' of research, first towards perspectives which acknowledged their subjectivity, albeit tempered by considerations of their cognitive abilities and social competence, and more recently to a greater recognition of children as social actors in their own right, inherently no less competent than adult respondents. A recent, related development is the growing interest among some – both practitioners and academics – in working with children and young people as co-investigators, on the basis that knowledge about children may be best produced by them. It was our engagement with this latter approach to knowledge production that encouraged us to propose the RGS/IBG session and this special issue, where co-production of knowledge is the key focus. The papers in this collection present a nuanced picture: they are not so much concerned with the question of whether young people can produce *better* knowledge than adults, but more with how to best understand the *different* knowledges that emerge from diverse actors within different generations and so ensure that maximum benefits accrue to the children and young people with and for whom the research is conducted.

In practicing participatory work with young people aimed at co-production of knowledge, contextual understanding – in particular, the impact of local social relations and the complexity of power inequalities (including both intra- and inter-generational relations) – is crucial. Not surprisingly then, context forms a pervasive theme across the papers in this collection, whether talking about Hull, Bratislava or a remote Lesotho village. Walker *et al.*'s study of the impact of Hull floods on children's experiences, for instance, considers the interconnections between pre-existing kinds of vulnerability and new forms of vulnerability associated with the uncertainties of secondary flooding on post-flood life: vulnerabilities which are both dynamic and contextual. The dangers of forgetting or downplaying embeddedness are also strongly foregrounded in the papers by Ansell *et al.* and Blazek and Hraňová. Both papers are concerned with how contextual *relations between young people* can shape the pattern and progress of a study. In the paper on participatory video in Bratislava (Blazek and Hraňová), the dynamics of pre-existing social relations among young people had to be incorporated in the participatory video development 'without disturbing its collective and collaborative nature'. The researchers, thus, found they had to work from the base of pre-existing power dynamics to *extend* the agency of the group, rather than trying to 'flatten' its power dynamic. In the very different social setting of rural Lesotho and Malawi, the research conducted by Ansell *et al.* suggests that the reluctance of young people to be open about their experiences in the emotional storyboard component of the research project may, in part, be related to pre-existing social relations among them: either children did not want to share difficult experiences and/or 'were following cultural taboos about discussing death'.

Another key element of context is the nature of *youth–adult relationships* and associated power relations. A vital but under-researched issue associated with adults working in apparently participatory ways with children and young people is the process whereby (at least seemingly) co-constructed knowledge emerges. This knowledge will be shaped by the nature of the relationship between young people and adult researchers and the extent to which young people as social agents are able to enter that equation. While youth–adult interaction is an underlying theme in all papers, this is most directly tackled in three (Todd, Ansell *et al.*, Jupp-Kina). The specific problematic of the way children's voices may be used by adult practitioners in UK children's services is strongly emphasized in Liz Todd's paper. Her concern is that the policy mandate to consult with children over decisions that are being made about them in diverse services has resulted in seemingly uncritical surveys of children's views by practitioners, tokenism and a failure to appreciate the heterogeneity of young voices. She suggests that being further up the (Hart 1992) participation ladder away from adult control is not always either necessary or better, given the political complexities that shape 'the production and reception of the child in research': marketization, responsibilization and disempowerment are real dangers such that 'recognition' of children's agency may still leave them merely legitimating policy. This leads to a call for critical dialogue and critical methodology to bridge the gap between practitioners and academics in research with children, with a more nuanced approach which looks critically at adult purpose and motivation, approaches to negotiation of consent, method (knowledge construction, not knowledge extraction) and interpretation (away from the uncritical treatment of voice). The need for (albeit gradual) change in the culture of UK professional practice is a conclusion which will resonate with many readers.

In Jupp-Kina's study of an NGO working with young people, supported by youth mediators (NGO staff aged 15–19 years), in Sao Paulo, Brazil, adult–youth relationships and the knowledges they produce are also centre-stage. The impact of emotions and power in participatory processes is explored (with particular reference to the youth mediators), drawing on the work of Spinoza and Foucault. The biggest challenge to adults in participatory practice identified here is not the inclusion of 'them', but taking on board issues of reflexivity: i.e. including 'ourselves' within the process, being prepared to face up to the inconsistencies of 'our own limits', accepting our own power and consequently accepting responsibility. Diverse *perceptions* of power and responsibility for change complicated matters in this context: young people (including young staff mediators) perceived adult power to make decisions as much greater than adult NGO staff themselves perceived that power. There was, thus, a sense of exclusion from decision-making processes amongst both young people and staff, leading to a passive acceptance of managerial decisions and consequent feelings of frustration. This illustrates how, while participatory processes require each person to understand their own particular pair of glasses, this can only occur if all those involved in the process wish to be transformed. As with Todd's paper, it suggests a need for practitioners to interrogate their professional practice and consider what fresh perspectives are needed in order to overcome blocks to knowledge flows.

In rural Lesotho and Malawi, Ansell *et al.* tackle the problem of adult mediation of children's voices from a slightly different angle since this is a research study without a direct practitioner element. In the paper they explore the different participatory methods which adults may use with young people in seeking to learn about children's experiences from children's own perspectives, but their interest is not just in how young people construct their experiences but how their experiences are constructed. The focus of their research is a particularly sensitive one – the impact of AIDS on livelihoods – and their experimentation with different methods in this context is illuminating. They explore the value of a number of different participatory methods in the co-production of knowledge with young people, and the contradictions between knowledges which may emerge, emphasizing the importance of taking fully into account the relationship between epistemology and methodology in selecting and employing methods appropriate to particular research questions. This leads them (like Todd), to conclude by urging caution in the uncritical use of participatory methods for empirical investigation. Ansell *et al.* also show the complexity of the insider–outsider position occupied by local research assistants (in one context themselves young, in the other context middle-aged, in both contexts interpreters from the same or nearby communities) and how this impacted on knowledge production. These assistants were given training, which drew on their own experiences, but they then tended to take this model to help the children shape their responses, in some cases to the extent of seemingly actually censoring children's input into the study!

The majority of work on young people's participation in knowledge production (including most work in this collection) is concerned with its immediate context, practice, impacts and associated ethical considerations. In this collection, however, a number of papers (Walker *et al.*, Blazek and Hraňová, Hampshire *et al.*) draw attention to issues of temporality. Walker *et al.* explore flood impacts on children's lives in the northern England city of Hull over a period of a year, emphasizing the importance of considering process rather than a mere snapshot view of events: new forms of vulnerability – some only evident through careful investigation – emerged over time amongst some of the participants. They emphasize that vulnerabilities cannot be reduced to a static list of criteria, because they are produced through the interaction between the recovery process and the specific circumstances at work in the children's lives. Their work on resilience, meanwhile, draws attention to the ways that children and young people have contributed to building family, community and ultimately city-wide resilience.

Blazek and Hraňová also take a temporal view, following the process of their participatory video project with young people through video training and production in Bratislava to showing of the video at an international youth exchange in Liverpool and the return of the young people with their video to Bratislava. They consider the way this project brought changes both within and beyond the group over time and influenced young people's subsequent involvement in the community. Hampshire *et al.* take an even longer view, looking to explore the impacts of children's research participation on their lives 2 years after their direct participation had ended. The research study of child mobility, conducted initially with 70 young researchers as co-investigators working alongside 'professional' academic researchers, in Ghana, Malawi and South Africa, enabled the young people involved to produce knowledge about their own worlds which then helped shape subsequent research by the adult academic team. While the inclusion of children as researchers in their own right may contribute to a fuller understanding of their perspectives on the issue under investigation, this can raise significant new ethical issues (Robson *et al.* 2009, Porter *et al.* 2010). Drawing on a small follow-up study in one country, Ghana, Hampshire *et al.* track forward to draw attention to issues of which, at the outset, we may not be able to take full cognizance: that children are changing, growing older, their relationships with adults are changing, and their lives unfold in sometimes unpredictable ways. The ethical implications are thus more complex and research design more difficult than we might initially anticipate.

To conclude, among both practitioners and 'professional' academic researchers concerned with children, the co-production of knowledge with children and young people is likely to expand and extend substantially, given the growth in effort and experimentation towards improved practices of participation. While the papers in this collection support that trajectory, they also reinforce the call for constant vigilance, not least the need to regularly revisit the question of ethics and to continually interrogate the inter- and intra-generational relationships within which knowledge is co-produced, reproduced and represented.

Gina Porter
Department of Anthropology, Durham University, UK

Janet Townsend
Department of Geography, Newcastle University, UK

Kate Hampshire
Department of Anthropology, Durham University, UK

References

Hart, R., 1992. *Children's participation: from tokenism to citizenship*. Florence: UNICEF.
James, A., 2007. Giving voice to children's voices: practices and problems, pitfalls and potentials. *American Anthropologist*, 109 (2), 261–272.
Porter, G., *et al.*, 2010. Children as research collaborators: issues and reflections from a mobility study in Sub-Saharan Africa. *American Journal of Community Psychology*, 46 (1), 215–227.
Robson, E., *et al.*, 2009. 'Doing it right?': working with young researchers in Malawi to investigate children, mobility and transport. *Children's Geographies*, 7 (4), 467–480.

'It came up to here': learning from children's flood narratives

Marion Walker[a], Rebecca Whittle[a], Will Medd[a], Kate Burningham[b], Jo Moran-Ellis[c] and Sue Tapsell[d]

[a]*Lancaster Environment Centre, Lancaster University, Lancaster LA1 4YQ, UK;* [b]*Department of Sociology and Centre for Environmental Strategy, University of Surrey, Guildford, UK;* [c]*Department of Sociology, University of Surrey, Guildford, UK;* [d]*Flood Hazard Research Centre, Middlesex University, London, UK*

> The growing body of literature that seeks to understand the social impacts of flooding has failed to recognise the value of children's knowledge. Working with a group of flood-affected children in Hull using a storyboard methodology, this paper argues that the children have specific flood experiences that need to be understood in their own right. In this paper, we consider the ways in which the disruption caused by the flood revealed and produced new – and sometimes hidden – vulnerabilities and forms of resilience and we reflect on the ways in which paying attention to children's perspectives enhances our understanding of resilience.

Introduction

Relatively few accounts of flooding and flood recovery take account of the perspectives and agency of children and young people. While there is a strong body of evidence that children are a vulnerable sub-group to flooding (Thrush *et al.* 2005a, 2005b), and some research has identified the need to understand children's perspectives on flooding (Tapsell 1997, Tapsell *et al.* 2001, RPA *et al.* 2004), most studies of natural hazards have failed to incorporate the growing body of research that recognises the role of children as social actors (Tucker and Matthews 2001, Valentine and Holloway 2002, Hemming 2008). This neglect is particularly problematic given the increasing policy emphasis on building individual and community resilience as a strategy for coping with floods (Defra 2005, 2008, Environment Agency 2005, Cabinet Office 2010). It is also problematic in the context of shifts in policy worlds, from the United Nations down to national and local government, that recognise the rights of children and young people to have a say in decisions that affect their lives (DCSF 2008).

This paper details some key findings of an in-depth study working with a group of flood-affected children and young people which set out to identify key aspects of their experiences and agency in relation to flooding and the flood recovery process. The research was based in the city of Kingston-upon-Hull, north east of England, which experienced severe flooding in June 2007. Over 110 mm of rain fell on the city during the biggest event on June 25th, overwhelming the drainage system and resulting in widespread pluvial flooding. The Hull floods affected over 8600 households and 1300 business properties, one young man died and 91 of the city's 99 schools were affected (Coulthard *et al.* 2007a, 2007b). However, our research shows that establishing who was affected – and how – is more complex than the statistics suggest.

To set the scene for the paper, we begin by reviewing the literature which discusses the impacts of flooding on children. We then turn to consider how these debates relate to issues of vulnerability and resilience, before moving on to discuss the methodology we employed in the study. In the paper, we show how the disruption caused by the flood revealed and produced new – and sometimes hidden – vulnerabilities and forms of resilience amongst the children. In conclusion, we reflect on the ways in which paying attention to children's perspectives can enhance our understanding of resilience and the role that children play in their communities more generally.

Children and flooding

The Pitt Review of the summer floods of 2007 in England (Cabinet Office 2008) supports the findings of an increasing body of social science literature that pays testament to the economic, social and emotional impacts of flood recovery (Tapsell *et al.* 2002, Fielding and Burningham 2005, Thrush *et al.* 2005a, 2005b, Walker *et al.* 2006). Nevertheless, children remain largely hidden from research on flood and flood recovery despite the fact that one in four households at risk of flooding in England and Wales have children living in them (Burningham *et al.* 2005). This omission is consistent with Valentine's (1997) claim that contemporary research on children's geographies reveals the extent to which adults know relatively little about children's social worlds.

To date, those studies which have attempted to explore the impacts of flooding on children have done so from an adult-focused perspective, rather than working with children themselves. Such work shows that children can be severely affected – both physically and emotionally – by natural disasters such as flooding (Flynn and Nelson 1998, Tapsell and Tunstall 2001). Studies have also highlighted the social and physical health effects of flooding upon children in a development context (Flynn and Nelson 1998, Delap 2000, Zoleta-Nantes 2002, Hossain and Kolsteren 2003). In the UK, research has shown that flood-affected children are prone to health problems such as coughs, colds and eczema (Tapsell *et al.* 1999, Tapsell and Tunstall 2001). Children may also experience emotional impacts; for example, parents interviewed in Carlisle reported that their children would still become upset and cry during heavy rain more than a year after the floods took place in 2005 (Watson *et al.* 2007). Parents identified childhood stress at home over the loss of possessions or pets and distress at school (Carroll *et al.* 2006, Convery *et al.* 2010), as well as behavioural problems, including difficulty sleeping, nightmares and tantrums (Welsh Consumer Council 1992, Hill and O'Brien 1999).

The flood recovery process can also impact adversely on children. One study cites poor academic performance as a result of stress in family relationships during the long-term flood recovery process (Allen and Rosse 1998), where such stress might come from the disruption to normal routines as well as social isolation from friendship networks (Tapsell and Tunstall 2001). In the north east of England, parents were angry that there was a lack of advice on how to deal with children after a flood, and that no social or psychological support had been provided for young people (Tapsell and Tunstall 2001). Nevertheless, parents may be too pre-occupied with 'crisis management' to consider how their children are affected (Ketteridge and Fordham 1995).

However, while we can draw inferences from such adult-focused studies, this does not give us systematic and robust insights into children's own experiences of flooding, or indeed how these experiences may relate to concepts of vulnerability and resilience which are frequently used to examine people's experiences of disaster recovery.

Flood, vulnerability and resilience

The concept of vulnerability has been an increasing concern within the literature that seeks to explore the social impacts of flooding (Tapsell *et al.* 2002, Fielding and Burningham 2005). Within this literature, vulnerability is often considered to be linked to particular demographic and socio-economic factors such as age, ethnicity, income, pre-existing poor health and family structure (Thrush *et al.* 2005a, 2005b, Walker *et al.* 2006). Following the logic of such arguments, children – like older people – may automatically be classed as a 'vulnerable group'. However, more recent research suggests that the situation is more complex than this. Working with a group of flood-affected adults in Hull, Whittle *et al.* (2010) suggest that vulnerability is both dynamic and contextual and that it cannot be reduced to a static list of socio-demographic characteristics that can be defined and measured, such as age or disability. Specific circumstances operating in a person's life (some of which were completely unrelated to flooding, such as redundancy or family illness) influenced who became vulnerable at different points during the recovery process. Such conclusions were born-out in our work in Hull where we discovered that some children may be vulnerable immediately before and after the flood, while others may become vulnerable as a result of the ways in which the long-term flood recovery process is played out (Walker *et al.* 2010, Whittle *et al.* 2012).

Whilst the concept of vulnerability focuses on weakness and susceptibility, resilience, in contrast, suggests a more positive sense of strength. Competing conceptions of resilience have proliferated across a wide range of literatures with different implications for what the analysis of building resilience might mean (Medd and Marvin 2005). In our study, we were interested in two manifestations of resilience: the extent to which resilience was already present and demonstrated in the flood response, and the extent to which new forms of resilience were being (or could be) established as a response to the flood. The role that children played as social actors was therefore important: concerns that children are a 'vulnerable' group has led to increased interest from both policy worlds and academia into the role that children can play in building resilience within their homes and communities (Cabinet Office 2010). By focusing on this issue we build on existing literatures on children's psychological resilience which discuss how children's services can enhance children's resilience (Resilience Research Centre Canada 2008), and which makes evaluations and suggestions for the development of education programmes for children and young people (Ronan and Johnston 2005; and in a development context Izadkhah and Hosseini 2005). The research also builds on a well-established literature that, since the 1990s, has demonstrated the importance of recognising the competencies and capacities of children as individual social actors who make sense of, and actively engage with, their social worlds (Valentine 1996, James and Prout 1997, James *et al.* 1998, Hutchby and Moran-Ellis 1998, Holloway and Valentine 2000, Smith and Barker 2001, Tapsell *et al.* 2001, James and James 2004, Newman *et al.* 2006).

In the project we explored the ways in which the children contributed to the process of building resilience (for example, by seeing children as having an active role in the recovery process (James and Prout 1997, Hutchby and Moran-Ellis 1998, Pain 2006). We also explored their role in the household – for example, by providing practical help as well as emotional support (Burningham *et al.* 2005, Seymour 2005, Thrush *et al.* 2005a, 2005b). We now outline the methods on which the study was based before going on to explore the children's narratives of the flood and the subsequent recovery process.

Methodology

The methodology builds on the growing body of work that uses interactive mixed-method research to generate rich data about issues in children's own lives (Thomas and O'Kane 2000, Tapsell *et al.* 2001, Morrow 2002, 2003, 2004, Ansell and van Blerk 2004, Kellett 2005,

Walker *et al.* 2009). In total, 46 flood-affected children took part, from May 2009 to May 2010; some of the children were flooded at school and others were flooded both at school and at home. Access to the children was via three sources: a primary school, a secondary school and youth groups. However, in this paper we concentrate solely on the experiences of the school groups, where the participants were aged 7–13 years when the floods occurred and 9–15 years at the point of data collection.[1] Hull City Council provided detailed information on how schools across Hull were affected and played a lead role in facilitating our relationship with the schools and youth groups. We worked in two schools that were badly affected by the floods; both schools were evacuated on the day and then closed for strip-out and refurbishment. The school children totalled 42 (25 primary and 17 secondary pupils) with Tables 1 and 2 providing a profile of the schools and the participants, respectively.[2] Both schools were also situated in areas characterised by high levels of social disadvantage: eligibility for Free School Meals is used as an indicator of poverty and Table 1 shows that the number of pupils entitled to free school meals is above the national average in both establishments.

Participating classes were chosen with staff guidance. We worked with the Year 5 class in the primary school (9–10-year-olds) and a mixed age-group of pupils from Years 7–10 (11–15-year-olds) in the secondary school.

Following recruitment, we used storyboards, interviews and group discussions with the school children and telephone interviews with four flood-affected young people, accessed through the youth team. We also interviewed 18 adults who supported children in Hull following the floods, including key service providers and front line workers, together with stakeholder engagement through a project steering group.[3]

The storyboard workshops involved 44 participants in two sessions; one for the primary school (16 children and three staff) and one for the secondary school (18 children and two staff). The primary school headteacher was concerned that some pupils had 'a nervousness' about rain and she suggested we talk about 'the fun things to do with water'. To generate a feeling of excitement, the workshops took place off the school premises and began with a set of games based around the theme of 'water'. The children were then asked to represent their

Table 1. School profiles.

School profile	Marshside Primary School	Edgetown Secondary School
Type	Community (LA maintained)	Community (LA maintained)
Admissions	NA	Comprehensive
Gender	Mixed	Mixed
Age range	3–11	11–16
Pupils	329	1200
Free school meals	178	312

Table 2. Project participants from the schools.

School participants	Marshside Primary School	Edgetown Secondary School
Cohort	26 (9–10 years)	17 (11–15 yrs)
Free school meals	15	1
SEN (pupils learning needs requiring extra support)	9	0
EAL (English as an additional language)	1	0

'flood journey' in a storyboard. Storyboards have been shown to be an effective method in children's research because they avoid problems associated with low literacy levels and give participants the freedom to 'represent' themselves in particular ways (Smith and Barker 2001, Ansell and van Blerk 2004, Newman et al. 2006, Hemming 2008). We then conducted one-to-one interviews with the children (15–20 min, 42 recorded in total – 25 primary and 17 secondary pupils) using the storyboards as prompts (Loizos 2000, Walker et al. 2009). For example, on his storyboard Zain (Figure 1) shows the flood water reaching up to, and then receding from, the letter box and we talked about this in the interview.

The audio-data from the interviews were transcribed in full and categorised thematically using data analysis software (Atlas Ti). This process involved coding of the data by the researchers, and data clinics in which the research team read a sample of the data and compared their interpretations in order to identify key themes for analysis. Issues around loss and disruption at home, at school and in friendship networks were prevalent within the children's narratives. As a result, the following sections use a framework of what was disrupted, revealed and created by the flood as a means to enhance our understandings of vulnerability and resilience.

The disruption of everyday life

Media coverage of flooding demonstrates only too clearly the immediate disruptions that floods can produce in the daily lives of a city and its inhabitants. The children described the immediate impacts of the flood in a number of ways. The most obvious of these was the effect that the flood had on their homes. The children had vivid recollections of the chaos caused as the floodwaters entered their homes, often in unexpected ways. Darren (Yr 5/10)[4] remembered, 'it started coming through the holes where you have the [tv] aerial, loads of it ... all over the house', while Wayne (Yr 5/10) noticed that the water went 'yellow' in the toilet and then water 'started squirting out' of the washing machine.

This initial experience was shocking for those who did not realise the extent and severity of the floods, as Gemma's (Yr 9/14) experience illustrates. Like many of the children, Gemma thought that there would only be a 'bit of damp', whereas the reality was much worse. Her

Figure 1. Zain's 'flood journey'.

bedroom was downstairs so she put 'all her stuff' on the top of her cupboard, took her uniform and slept upstairs. However, what happened next came as a shock:

> At 4 o'clock in the morning the next thing I know the house was flooded ... I didn't know my house was going to be flooded. I thought it was just going to be like a little bit of damp ... it was really high. Because where I live it's like a bowl and all the water just came in and they were all sucking in the water to get it out, the fire brigade and that lot, it was coming back down our street, it was really high. (Gemma Yr 9/14)

The children's accounts then show how this initial disruption was followed by the more hidden but no less troublesome disruption involved in the longer term recovery process. Here, we can see how the recovery process affected the daily routines and social practices of the household – a point which was well illustrated by the experiences of those who lived in caravans during the repairs to their home. When talking about life in the caravan, it was common for the children to use words such as 'cramped' or 'squashed', and this meant that everyday activities had to be reconfigured to adjust to the lack of space and facilities. Sally (Yr 7/11) illustrates how her family changed their cooking and eating habits in the aftermath of the floods. To begin with, she said: 'we got takeaways but then we stopped and we was having salads because it was too unhealthy'. Then her parents bought a small touring caravan and they put it on the drive way. Sally said 'my mum started cooking ... and some days we had like meals like spaghetti bolognaise and that and hot dinners ... but it wasn't very good, well it tasted nice but the cooker wasn't good'. Sally and her sister helped with the washing up in the caravan 'but it was only like a little sink and the fridge was really small so we had another fridge in the garage and a freezer'.

Socialising, doing homework and even sleeping were difficult in such close quarters and, in many cases, this disruption took its toll on the physical and emotional well-being of the children and their families, as Megan (Yr 9/14), whose family of six lived in a four-berth touring caravan, describes in this extract from her storyboard:

> ... when me and my little sister came back we had to live in the tiny caravan because my littlest sister had just turned 2, all she would do is cry. So none of us would sleep. My little sister was still crying every night so we took her to the doctors. My little sister had pneumonia so the builders had to work really fast so my little sister could move out of the caravan. When Christmas came we had to spend it at my grandma's so there was 7 of us in a one bedroomed flat! And my grandma had a close friend over so there was 8 of us! But it wasn't too bad. And after a week of staying there we went home and got back in our house. So we stayed in a caravan for 4 months! (extract taken from Megan's storyboard).

Megan's experience also illustrates how such disruption did not only affect the daily routines of the household: instead, its significance became particularly apparent during the longer term rhythms of family life – including events such as Christmas and birthdays. For example, Victoria's (Yr 8/12) parents were separated, yet both their houses were affected by the flood. Her dad's house 'wasn't finished or anything for Christmas ... it was really small so ... not much space to do all Christmas stuff'. Added to this Victoria's mum not having insurance and money became an issue. She said, 'my mum couldn't really afford much for Christmas because we had to do the floor and other stuff to like the house and it like wrecked our telly as well'. We return to Victoria's story in the next section where we discuss the kinds of vulnerabilities revealed by the flood.

However, it is important to note that the disruption experienced by the children did not stop at the threshold of the home – it also extended to their gardens and the streets and parks where they played and hung out, as Figure 2, with the image of a miserable looking boy playing football and riding his bicycle through the puddles, shows.

Figure 2. Playing outside is not easy.

Sherry's (Yr 5/10) mum told her not to play in the water because 'it was all contaminated' and although she didn't know what it meant at the time in the interview she said 'I do now . . . It's like all dirty, it's got muck in it, it's like the drains'. Gemma (Yr 9/14) wouldn't go near the 'rank' water, saying 'I wouldn't touch it . . . it was horrible, it was all brown'. Tim's (Yr 5/9) storyboard (Figure 3) shows 'the poo' floating past his front door.

Such changes to outside spaces also disrupted the children's ability to socialise with their friends and enjoy some independence away from their homes. This was particularly the case

Figure 3. The 'poo' in the brown water.

for young people like Josh (Yr 7/12) who moved to rented housing in another part of the city while his home was repaired. Being in rented housing meant that Josh could no longer walk to school with his friends or hang out with them as usual – if he wanted to spend time with them he was dependent on his father dropping him off in the car and collecting him later.

The final kind of disruption to highlight was that experienced at school. Understanding this kind of disruption is vital as 91 of Hull's 99 schools were affected (Coulthard et al. 2007b) and, as a result, many children and young people were affected both at home and at school. This disruption was particularly pronounced for the younger children at Marshside who had to attend two different schools while their own school was repaired. Getting used to a different space, a new school journey and the co-presence of other children was an unsettling experience for some. Charlie (Yr 5/10) said 'It was a bit scary because I've never been to a different school before'. The first school the children attended was the local secondary school, and this was an issue for him because the pupils at the comprehensive were so much older and bigger. Charlie said, 'when we like went for dinner the ... kids were there having their dinner and they kept swearing and all that to us'. Then the children moved schools again, in September 2007 to a primary school away from their estate. The disruption involved in the move was especially pronounced because the estate on which Marshside is located is very territorial and moving off the estate – even to go to the city centre – was not the norm for the children taking part in our project. The importance of exploring these contextual factors when attempting to understand children's experiences of disaster recovery is a theme that we return to in the discussion.

Revealing hidden vulnerabilities

Victoria's example illustrates how the flood revealed and, in some cases, exacerbated existing vulnerabilities. As explained previously, Victoria lived between her mother and father's separate homes, both of which were flooded. At her mum's house she said 'It's [the mould] going up the walls ... you can see the rising damp'. But because her mum is an unemployed owner occupier without insurance who does not have enough savings to pay for the repairs Victoria is not sure if the problem with the damp will be resolved, 'we don't know whether we are going to be moving house or something but if we aren't it will probably just stay like that'. Her mum's house is 'real cold and it's got a weird smell' and she said, 'I prefer being at my dad's new house' because it is 'bigger' than the one he lived in before the floods. Nevertheless en route to the bigger house Victoria's dad moved twice, 'he moved into a new house and they didn't realise it was flooded until my dad had been there and decorated and everything. So once they found out it was flooded my dad had to move again'. So he moved to another house and 'they've just found out that was flooded so they are going to have to live upstairs now'. As we discussed earlier in relation to the issue of Christmas presents, Victoria was conscious that the floods had cost both her parents money. Even though her dad rents Victoria said, 'he hasn't got as much money no more because he has to like do the house up and things'. However, the disruption her dad experienced also 'cost' Victoria because she had less time to spend with her dad.

Victoria's account of her dad's experiences living in rented accommodation is consistent with the findings of the adults' project where we discovered that private renters were particularly vulnerable during the longer term recovery process due to rising rents caused by a rapid increase in the level of demand for alternative rented housing across the city. Renters also had little or no control over their repairs as matters were handled by their landlords (Whittle et al. 2010).

Parallel research conducted with the adults in Hull showed that the impacts of the flood went much wider than the physical spread of the flood waters – for example, with serious effects on the relatives and carers of flooded residents (Sims et al. 2009). The children's accounts of the flood and the subsequent recovery process also reveal the spatial and temporal complexity of understanding who was affected and how. This is something that we have already alluded to in the previous section where we examined the disruption experienced by the children at school. Given that 91 of 99 schools in the city were affected, we can see how those whose homes were not affected could still have experienced impacts. However, the patterns of impact and, consequently, of vulnerability and resilience were more complex still. This complexity was illustrated by the experiences of Bob and Natalie.

Natalie (Yr 7/11) was not flooded in June 2007. However, her dad and step mum bought a new house in December 2007, which they believed to have been unaffected by the floods. Shortly after moving in, however, 'we started seeing all the damp up the walls and in the back room, all the weird brown dots and it started sticking and everyone got ill... Everyone started getting headaches and being sick'. Her dad and step mum then became embroiled in a protracted argument with the bank and the surveyors about the cause of the damage and – crucially – who should pay for the repairs. Natalie drew a black coloured bank and the words 'fight with bank for 8 months', a drawing of her step mum with a red, angry-looking face beside the words 'It's got rising damp' and 'it's not in writing it's floods' attributed to 'bank person' and then another of her step mum beside the word 'stressed' on her storyboard (Figure 4). The result of this dispute was that Natalie and her family were still living in rented accommodation 2 years after the flood while they waited for the problem to be sorted.

Natalie's story thus reveals the hidden stress that 'secondary flooding' imposed on children and their families, as well as the far-reaching implications of the flood in time and space. In contrast to 'official' accounts of disaster (which confine the incident to a particular place and time), we can see how the impacts are more complex and go wider in their reach (Walker et al. 2011). Her story also reveals the social processes – in terms of the interactions with the bank and the surveyors – that lie behind the designation of 'flood victim' status.

In contrast to Natalie, whose family was trying to convince the authorities that they had been affected, Bob (Yr 5/10) explained that it was not flooded. He gave us this explanation because he

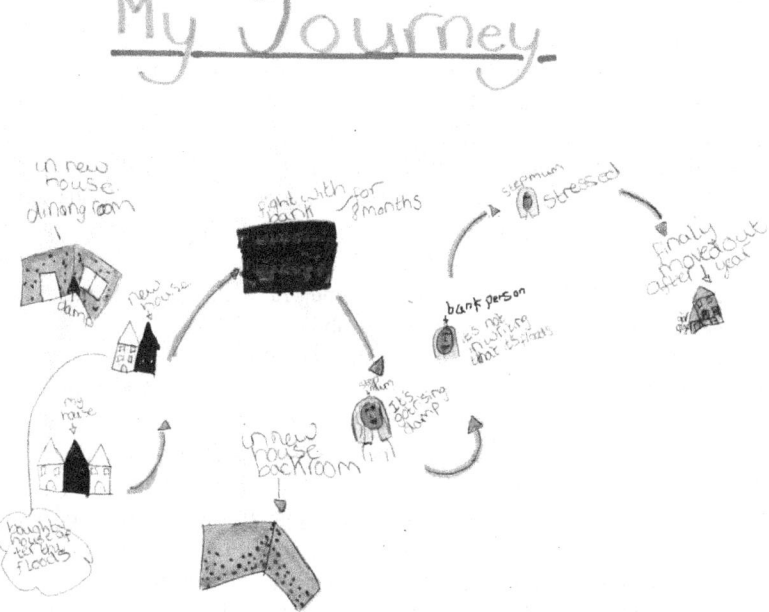

Figure 4. The stress.

officially 'lives' with his mum and her house was untouched. However, on a closer inspection, Bob revealed that he actually slept at, and spent a huge amount of time at, his dad's house, which was flooded. He explained:

> I go home and then have my tea and go out for a bit and I sleep at my dad's; I go to my dad's when I've larked out for a bit'. Most of the nights I stay there ... he goes to work early, about five, he starts at six but I stay in bed and walk home in the morning ... and then I get dressed and that at mine and then I come to school. (Bob Yr 5/10)

The flood damage to his dad's house meant that he could no longer stay over and he missed his dad as a result. Like Natalie, Bob's example illustrates the hidden impacts of the flood as well as the importance of understanding the children's experiences in the context of their everyday lives.

Producing new vulnerabilities and forms of resilience

In addition to disrupting and revealing particular aspects of the children's lives, the flood also had a more active role in producing new situations and challenges for the participants. One of the most fundamental experiences produced by the flood was the experience of loss. Many of these losses were only too evident in the physical damage created by the floodwaters. Cheryl was upset that she had lost her treasured dolls house 'forever', while Gemma lost sentimental things, including her jewellery box, her diary and her photographs because her bedroom was on the ground floor,

> The worst thing was probably my bedroom getting messed up because all my stuff was in there and the only thing that got rescued was my teddies because they were on top of my bed; they had like a bed up there and then like loads of storage space.

She said she lost things in the clean-up process when everything was thrown into the skip, 'I lost my jewellery because that was something they chucked it away, and I forgot all my jewellery was in it so all my jewellery got lost'. Others spoke of the sadness at losing pets: Victoria's rabbit lived in a hutch in the garage and he died before the family could rescue him.

However, our research shows that it is also important to be alert to the kinds of losses that are less immediately obvious. In particular, most of the children felt that they had lost valuable time with friends and family – for example, Josh, who moved to a rented accommodation away from his friends, and Bob who was not able to see his Dad as much.

Here, it is useful to return to our original concepts of vulnerability and resilience. On a more fundamental level we can see how the flood not only revealed pre-existing kinds of vulnerability (as was the case for Victoria living between two homes with both parents coping with financial hardship and whose experience we discussed in the previous section); it also produced new forms of vulnerability amongst some of the participants. Some of these vulnerabilities, such as those of Megan's little sister in the caravan, were only too obvious. However, others were much more subtle and only become apparent through a detailed exploration of the children's accounts. Here, we think particularly of Natalie, whose family life was overshadowed by the uncertainty of 'secondary flooding' 2 years later, and Bob who 'wasn't flooded' but found it hard to see his father. Such accounts are important because they support research which suggests that vulnerability is both dynamic and contextual (Walker *et al.* 2010). Looking at the detail of the children's accounts, we can see how, although vulnerability may sometimes be associated with particular characteristics – such as private rental households or those without insurance, this is not a straightforward picture. Indeed, vulnerabilities cannot be reduced to a static list of socio-economic or demographic criteria because vulnerability is produced through the ways in which the recovery process is played out in relation to the specific circumstances at work in the children's lives.

However, there was also some evidence that the flood may have produced particular forms of resilience. As we discussed previously, many of the children experienced disruption as the floodwater entered their homes and streets. However, this was not by any means an entirely negative experience for all of the participants, and neither were they passive observers of this process. For example, Josh explained that he was outside with his dad, trying to empty the water out of their garden with dustbins, while Michael helped his family carry furniture and other items upstairs. For some, this process of helping out continued into the repairs process where some of the children reported making tea for – and cleaning up after – the builders. Outside the home, too, many of the children reported fun and excitement and, although some of them stayed out of the water as a result of their own (or their parents') fears about contamination, others enjoyed playing in it.

Darren (Yr 5/10) said that the best thing 'was that I could sit on the balcony and catch fish' and Hayley (Yr 7/12) also went fishing, 'all the ponds got flooded and they were all like swimming about in the water'. Such examples illustrate how most of the children employed a creative use of agency. For example, Hayley also helped her parents by taking care of her 2-year-old brother who wanted to go fishing but was too small to go on his own. As Hayley explained 'He couldn't walk for the water' so she carried him, 'I said, "just sit on my back and I'll take you"'. Her ability to appreciate a more positive side of the floods was apparent from the way in which she divided her storyboard into 'high and low' points (Figure 5).

If we think of resilience as the ability to recover successfully, then the children's comments on their newly refurbished school are also an important indication. Although many of them found attending a temporary school to be stressful, there were plus points – Cheryl and some of the others liked the fact that there was a new bus journey to enjoy – and, when they eventually returned to their original school, they 'felt a lot more happier with the design and everything

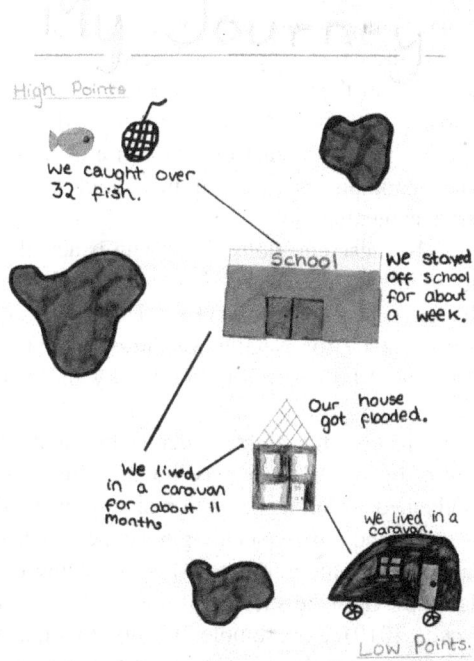

Figure 5. The highs and lows of Hayley's journey.

else ... There were new toys and all that ... New carpets, new toys, new TV, new books, new everything ...'.

Finally, resilience also emerged in unexpected ways. Interestingly, the complex family lives led by many of the children proved to be an unexpected source of resilience when the floods occurred. For those used to moving locally between the homes of different family members, the additional disruption caused by the flood requiring 'multiple moves' did not seem as great as it did to those with more settled home circumstances. There are, however, two points to set against these arguments: first, the moves made by the children took place in a relatively small spatial framework (i.e. within Hull), while moving beyond this was more problematic. Second, it is important to consider Erikson's (1976, 1994) comments about the kinds of 'pre-existing disasters' that may exist in poor communities. There was no doubt that many of the children and young people we worked with came from backgrounds where poverty – and the kinds of social hardships that can result from this – was a real problem for their families on a day-to-day basis. In a sense, then, the flood and the subsequent recovery process made these pre-existing problems more visible (as in the case of Victoria) as well as exacerbating their impacts on the community (Pelling 2003, Gunewardena 2008).

Discussion

This paper details the findings of research that set out to identify key issues in children and young people's experiences in relation to vulnerability and resilience to flooding and the flood recovery process.

As a first step we have shown that it is important to understand the social impacts of flood from the perspectives of flood-affected children themselves, rather than trying to make inferences about their issues from the accounts of adults. This is vital because children can and do define their vulnerability differently to the concerns that adults may have for them. For example, the adults we interviewed were particularly concerned about the impact of the floods on the children's examination results or on their ability to complete homework whilst living in caravans (Walker et al. 2010). However, as this paper shows, the young people themselves were often more concerned about the daily disruption to their lives – for example, the anxiety involved in having 'strangers' in your home, losing time with family members and an inability to socialise with their friends. There is no doubt that the impact of living in a caravan or living upstairs during the renovations, indeed the changing spatiality of the children's home-life, impacted upon family dynamics and hence is contextual.

The children's narratives also raise questions about who is actually affected by a disaster – and how. For some of the children, like Megan, it was only too obvious that they had been flooded and that they had experienced impacts as a result. However, in other cases this distinction was less than clear cut; life at home for Natalie was stressful living with the uncertainty of knowing whether her home would be classed as flooded by the insurance company (Walker et al. 2011).

There are also differences in how the children define flooding. Bob said he 'hadn't been flooded' because he thought of his Mum's house as 'home'. However, his Dad's house (where he slept most nights) was flooded, resulting in major disruption to Bob's life. Consequently, without speaking to Bob – and without understanding his experiences in the context of the complexity of his family circumstances – the impacts on his life would have been overlooked.

Children's accounts also show how the stresses of flood recovery have consequences for the whole family (see Whittle et al. 2010). For example, we might look at Victoria's mum and think that the financial stresses resulting from the damage to her home are personal to her – which of course they are, in one sense. However, Victoria's story shows how the whole family's life was affected in a much deeper way – Christmas was a sadder, sparser affair and Victoria did not enjoy being in her mum's house anymore because of the mould and the damp. She also really empathized with her mum's anxiety about the repairs and, of course, she was equally concerned about her dad. Although pre-existing vulnerabilities can be important (as we saw in the case of Victoria's family), the children's accounts have shown how the impacts of the flood are, in reality, much more subtle and wide ranging as vulnerability is produced through the ways in which the recovery process interacts with the children's circumstances and daily lives.

There is more, though, to this argument. There is a risk that we focus here on the children as 'victims' of the flood, valuing their voices, but not taking seriously their role as social agents. This would be to neglect the established literature that, since the 1990s, has demonstrated the importance of recognising the competencies and capacities of children and young people as individual social actors who make sense of, and actively engage with, their social worlds (Valentine 1996, James and Prout 1997, Hutchby and Moran-Ellis 1998, James et al. 1998, Holloway and Valentine 2000, Smith and Barker 2001, James and James 2004, Newman et al. 2006). As with other social actors, children and young people's agency is seen to arise from their structural and ideological position in society (Matthews and Limb 1999). Interestingly, such positioning can arguably be seen to have shifted with the UN Convention on Rights of the Child (1989), ratified by the UK Government in 1994, and, at a policy level, implemented in the UK's strategy of Every Child Matters in which 'children and young people will have far more say about issues that affect them as individuals and collectively' (DCSF 2008).

By bringing this recognition of children and young people's *agency* – as well as their rights to have a voice – to debates on building resilience, it is important to ask how children and young people can contribute to building future community resilience while at the same time examining how such a contribution is inhibited or enhanced by forms of institutional support. For example, by seeing children as agentic, we can explore the ways in which the positive coping and survival strategies illustrated in their accounts above (cf. James and Prout 1997, Hutchby and Moran-Ellis 1998, Pain 2006) contribute to family, community and, ultimately, city-wide resilience. Here, we think of Hayley – in taking her younger brother out fishing and distracting him from some of the more upsetting aspects of the flood, she was enabling her parents to concentrate on managing the immediate practicalities within the home. We can also note examples of children playing a role in the resilience of the household – perhaps by offering a source of physical support, comfort, practical help and a reason for 'carrying-on' (Burningham et al. 2005, Thrush et al. 2005a, 2005b). For example, we think of Josh and Michael, who helped their parents bale out water and rescue furniture, as well as those who made tea and cleaned up after the builders.

Conclusion

In this paper we have concentrated on a case study of flooding. However, the arguments explored here also address broader issues around disaster and recovery more generally. The children's accounts reveal that the impacts were both spatial and temporal and that these dimensions reinforce the need to move from a snapshot view of the reactions to natural disasters to a more processual one.

Disasters can be moments of transformation as the existing physical and social infrastructure appears to be swept away. The recovery process is therefore an important time as decisions must be made about how such infrastructure is replaced and, crucially, whose interests are represented in this process (Pelling and Dill 2010). In particular, Gunewardena (2008) argues that the kinds of policies and practices put in place after disasters should be targeted at reducing the inequalities that made local people vulnerable to the disaster in the first place (Pelling 2003). Much of this literature comes from a developing world context. However, its conclusions have equal relevance to case studies from the UK: given the current academic and policy interest in community resilience (Cabinet Office 2010) there is a need to explore the roles that children and young people play in the recovery process. Specifically, we can investigate how they may bring together community networks through their schooling, leisure and friendship networks (Ronan and Johnston 2005), whether children helping their family, their siblings and the wider community during the flood-recovery process also helps them and of course how their experiences of the flood as children might impact on their role in community resilience in adult life. There is also a need to build on existing literatures on children's psychological resilience with discussions of how children's services could enhance children's resilience (Resilience Research Centre Canada 2008) to a range of hazards and challenges in future. Doing so could help identify the ways in which policy could contribute to children's resilience, as well as to developing more enabling and empowering strategies which recognise the roles that children can play within their communities more broadly.

Acknowledgements

We wish to thank all of the children and young people who contributed to the project; they have been our inspiration throughout this research. Our thanks also go to the key service providers, front line workers and staff at the participating schools and youth groups for their co-operation in helping us to conduct the research. The project was funded by the Economic and Social Research Council, Environment Agency and Hull City Council.

Notes

1. See the project report (Walker *et al.* 2010) for more details regarding the methodology.
2. To maintain anonymity, pseudonyms have been used for both the schools and the participants.
3. The project steering group comprised local and national organisations with an interest in children's welfare and flood recovery: Lancaster University, the University of Surrey, Middlesex University, Hull City Council, the Environment Agency and representatives from the participating schools. The steering group was involved at all stages of the project from the original phases of designing the methodology through to later phases of analysis and dissemination.
4. School year and participant's age provided at the time of the interview.

References

Allen, R. and Rosse, W., 1998. *Children's response to exposure to traumatic events*. Boulder, CO: Colorado Hazards Centre.
Ansell, N. and van Blerk, L., 2004. Children's migration as a household/family strategy: coping with AIDS in Lesotho and Malawi. *Journal of Southern African studies*, 30 (3), 673–690.
Burningham, K., Fielding, J., and Thrush, D., 2005. *Public response to flood warning*. Report to the Environment Agency.
Cabinet Office, 2008. *The Pitt review: lessons learned from the 2007 floods*. London: The Cabinet Office.
Cabinet Office, 2010. *Strategic national framework for community resilience* [online]. London, The Cabinet Office. Available from: http://www.cabinetoffice.gov.uk/media/349129/draft-snframework.pdf [Accessed 5 October 2010].
Carroll, B., *et al.*, 2006. *Living in fear: health and social impacts of the floods in Carlisle 2005*. Research Report, University of Cumbria.
Convery, I., Balogh, R., and Carroll, B., 2010. Getting the kids back to school: education and the emotional geographies of the 2007 Hull floods. *Journal of flood risk management*, 3 (2), 99–111.
Coulthard, T., *et al.*, 2007a. *The June 2007 floods in Hull: interim report by the Independent Review Body Kingston-upon-Hull*. UK Independent Review Body.
Coulthard, T., *et al.*, 2007b. *The June 2007 floods in Hull: final report by the Independent Review Body Kingston-upon-Hull*. UK Independent Review Body.
DCSF, 2008. *Every child matters* [online]. Available from: http://www.everychildmatters.gov.uk/ [Accessed 23 July 2008].

Defra, 2005. *Making space for water: developing a new government strategy for flood and coastal erosion risk management in England: a delivery plan*. London: Defra.

Defra, 2008. *Future water: the government's water strategy for England*. London: Defra.

Delap, E., 2000. Urban children's work during and after the 1998 floods in Bangladesh. *Development in practice*, 10 (5), 663–673.

Erikson, K., 1976. *Everything in its path: destruction of community in the Buffalo Creek flood*. New York: Simon and Schuster.

Erikson, K., 1994. *A new species of trouble: explorations in disaster, trauma, and community*. New York: Norton and Co.

Fielding, J. and Burningham, K., 2005. Environmental inequality and flood hazard. *Local environment*, 10 (4), 379–395.

Flynn, B.W. and Nelson, M.E., 1998. Understanding the needs of children following large-scale disasters and the role of government. *Child and adolescent psychiatric clinics of North America*, 7 (1), 211–230.

Gunewardena, N., 2008. Human security versus neoliberal approaches to disaster recovery. *In*: N. Gunewardena, M. Schuller, and A. De Waal, eds. *Capitalizing on catastrophe: neoliberal strategies in disaster reconstruction*. Plymouth: AltaMira Press, 3–16.

Hemming, P., 2008. Mixing qualitative research methods in children's geographies. *Area*, 40 (2), 152–162.

Hill, J. and O'Brien, P., 1999. *Disaster in the community: emergency planning for sustainable solutions to long-term problems*. Worker and resident perspectives of the North Wales floods 1990 and 1993 Caernarfon, Gwynedd Disaster Recovery and Research Team Ltd.

Holloway, S. and Valentine, G., 2000. Spatiality and the new social studies of childhood. *Sociology*, 34 (4), 763–783.

Hossain, S.M. and Kolsteren, P., 2003. The 1998 flood in Bangladesh; is different targeting needed during emergency and recovery to tackle malnutrition? *Disasters*, 27 (2), 172–184.

Hutchby, I. and Moran-Ellis, J., eds., 1998. *Children and social competence*. London: Falmer.

Izadkhah, Y. and Hosseini, M., 2005. Towards resilient communities in developing countries through education of children for disaster preparedness. *International journal of emergency management*, 2 (3), 138–148.

James, A. and James, A., 2004. *Constructing childhood: theory, policy and social practice*. Basingstoke: Palgrave Macmillan.

James, A. and Prout, A., 1997. *Constructing and reconstructing childhood*. 2nd ed. London: Falmer.

James, A., Jenks, C., and Prout, A., 1998. *Theorising childhood*. Cambridge: Polity Press.

Kellett, B., 2005. *How to develop children as researchers: a step by step guide to teaching the research process*. London: Sage.

Ketteridge, A.M. and Fordham, M., 1995. Flood warning and the local community context. *In*: J.W. Handmer, ed. *Flood warning: issues and practice in total system design*. Enfield: Flood Hazard Research Centre, Middlesex University, 189–199.

Loizos, P., 2000. Video, film and photographs as research documents. *In*: M.W. Bauer and G. Gaskell, eds. *Qualitative researching with text, image and sound*. London: Sage, 93–107.

Matthews, H. and Limb, M., 1999. Defining an agenda for the geography of children: review and prospect. *Progress in human geography*, 23 (1), 61–90.

Medd, W. and Marvin, S., 2005. From the politics of urgency to the governance of preparedness: a research agenda on urban vulnerability. *Journal of contingencies and crisis management*, 13 (2), 44–49.

Morrow, V., 2002. 'They don't ask us': taking account of children's views in improving local environments. *In*: K. White, ed. *NCVCCO (National Council for Voluntary Child Care Organisations). Annual Review Journal*, 3, 51–66.

Morrow, V., 2003. Perspectives on children's agency within families: a view from the sociology of childhood. *In*: L. Kuczynski, ed. *Handbook of dynamics in parent–child relations*. Thousand Oaks, CA: Sage Publications, 109–130.

Morrow, V., 2004. Networks and neighbourhoods: children's accounts of friendship, family and place. *In*: C. Phillipson, G. Allan, and D. Morgan, eds. *Social networks and social exclusion: sociological and policy issues*. Aldershot: Ashgate, 50–71.

Newman, M., Woodcock, A., and Dunham, P., 2006. 'Playtime in the borderlands': children's representations of school, gender and bullying through photographs and interviews. *Children's geographies*, 4 (1), 289–302.

Pain, R., 2006. Paranoid parenting? Rematerializing risk and fear for children. *Social and cultural geography*, 7 (2), 221–243.

Pelling, M., 2003. *The vulnerability of cities*. London: Earthscan.

Pelling, M. and Dill, K., 2010. Disaster politics: tipping points for change in the adaptation of sociopolitical regimes. *Progress in human geography*, 34 (1), 21–37.

Resilience Research Centre Canada, 2008. http://www.resilienceresearch.org [Accessed 23 July 2008].
Ronan, K.R. and Johnston, D.M., 2005. *Promoting community resilience in disasters: the role of schools, youth, and families*. New York: Springer.
RPA, FHRC, EFTEC & CASPAR, 2004. *The appraisal of human-related intangible impacts of flooding*. Report to Defra/Environment Agency. R and D Project FD2005. London: Defra. http://www.rpaltd.co.uk/documents/J359-FD2005-Final2.pdf [Accessed 9 March 2012].
Seymour, J., 2005. Entertaining guests or entertaining the guests: children's emotional labour in hotels, pubs and boarding houses. *In*: J. Goddard, S. McNamee, A. James, and A. James, eds. *The politics of childhood: international perspectives, contemporary developments*. London: Palgrave Macmillan, 90–106.
Sims, R., *et al*., 2009. When a "Home" becomes a "House": care and caring in the flood recovery process. *Space and culture*, 12 (3), 303–316.
Smith, F. and Barker, J., 2001. Commodifying the countryside: the impact of out-of-school care on rural landscapes of children's play. *Area*, 33 (2), 169–176.
Tapsell, S.M., 1997. Rivers and river restoration: a child's-eye view. *Landscape research*, 22 (1), 45–65.
Tapsell, S. and Tunstall, S., 2001. *The health and social effects of the June 2000 flooding in the North East region*. Report to the Environment Agency. Flood Hazard Research Centre, Middlesex University.
Tapsell, S., *et al*., 1999. *The health effects of the 1998 Easter flooding in Banbury and Kidlington*. Report to the Environment Agency, Thames Region Flood Hazard Research Centre, Middlesex University.
Tapsell, S., *et al*., 2001. Growing up with rivers? Rivers in London children's worlds. *Area*, 33 (2), 177–189.
Tapsell, S., *et al*., 2002. Vulnerability to flooding: health and social dimensions. *Philosophical transactions of the Royal Society*, 360 (1796), 1511–1525.
Thomas, N. and O'Kane, C., 2000. Discovering what children think: connections between research and practice. *British journal of social work*, 30 (6), 819–835.
Thrush, D., Burningham, K., and Fielding, J., 2005a. *Exploring flood-related vulnerability: a qualitative study*. R and D Report W5C-018/3. Bristol: Environment Agency.
Thrush, D., Burningham, K., and Fielding, J., 2005b. *Vulnerability with regard to flood warning and flood event: a review of the literature*. R and D Report W5C-018/1. Bristol: Environment Agency.
Tucker, F. and Matthews, H., 2001. 'They don't like girls hanging around there': conflicts over recreational space in rural Northamptonshire. *Area*, 33 (2), 161–168.
Valentine, G., 1996. Angels and devils: moral landscapes of childhood. *Environment and planning D: society and space*, 14 (5), 581–599.
Valentine, G., 1997. 'Oh yes I can.' 'Oh no you can't' Children and parents' understandings of kids' competence to negotiate public space safely. *Antipode*, 29 (1), 65–89.
Valentine, G. and Holloway, S., 2002. Cyberkids? Exploring children's identities and social networks in on-line and off-line worlds. *Annals of the association of American geographers*, 92 (2), 302–319.
Walker, G., *et al*., 2006. *Addressing environmental inequalities: flood risk*. Science Report SC020061/SR1. Bristol: Environment Agency.
Walker, M., *et al*., 2009. Talk, technologies and teenagers: understanding the school journey using a mixed-methods approach. *Children's geographies*, 7 (2), 107–122.
Walker, M., *et al*., 2010. *After the rain has gone: learning lessons about flood recovery and resilience from children and young people in Hull*. Final project report for Children, Flood and Urban Resilience: Understanding children and young people's experience and agency in the flood recovery process. Lancaster, UK: Lancaster University.
Walker, G., *et al*., 2011. Assembling the flood: producing spaces of bad water in the city of Hull. *Environment and planning a special edition*, 43 (10), 2304–2320.
Watson, N., *et al*., 2007. *Understanding response and resilience in post-flood communities: lessons from the Carlisle pilot study*. Lancaster, UK: Lancaster University.
Welsh Consumer Council, 1992. *In deep water: a study of consumer problems in Towyn and Kinmell Bay after the 1990 floods*. Cardiff Welsh Consumer Council.
Whittle, R., *et al*., 2010. *After the rain – learning the lessons from flood recovery in Hull*. Final project report for Flood, Vulnerability and Urban Resilience: a real-time study of local recovery following the floods of June 2007 in Hull. Lancaster UK: Lancaster University.
Whittle, R., *et al*., 2012. Flood of emotions: emotional work and long-term disaster recovery. *Emotion, Space and Society*, 5 (1), 60–69.
Zoleta-Nantes, D.B., 2002. Differential impacts of flood hazards among the street children, the urban poor and residents of wealthy neighbourhoods in Metro Manila, Philippines. *Mitigation and adaption strategies for global change*, 7 (3), 239–266.

Emerging relationships and diverse motivations and benefits in participatory video with young people

Matej Blazek[a] and Petra Hraňová[b]

[a]School of the Environment, University of Dundee, Dundee, UK; [b]Civic Association Ulita, Bratislava, Slovakia

> The paper reflects on the process of participatory video production with young people from a deprived neighbourhood in Bratislava. We draw on Kindon's [2003. Participatory video in geographic research: a feminist practice of looking? *Area*, 35 (2), 142–153] and Parr's [2007. Collaborative film-making as process, method and text in mental health research. *Cultural geographies*, 14 (1), 114–138] arguments that the *process* of participatory video can bear more significance for all actors of the video than the video-as-a-product. The paper thus explores relationships between particular groups of actors (young participants, the researcher and the practitioner) as well as among them, in the video-making process. We are especially interested in the diversity of motivations behind different actors' decisions to be involved in participatory video, and we explore the dynamic changes of such motivations and the range of ultimate benefits that participatory video provided. These insights in turn help us to understand multiple types and layers of knowledge produced by young people through participatory video. We conclude the paper by highlighting the intersubjective diversity of participatory video, and we suggest how this can be approached to make participatory video research transformative and efficient for the purpose of research at the same time.

Introduction

In her benchmark paper, Kindon (2003) introduced participatory video as an accessible yet efficient tool for geographical research. Kindon argued that participatory video has the potential to challenge 'conventional relationships of power' (p. 143) characteristic for geographical fieldwork, especially those between researchers and participants from disadvantaged social groups. Since its publication, Kindon's paper, (later expanded also elsewhere, see for instance Hume-Cook *et al.* 2007, Kindon 2009) has attracted attention within the discipline, especially among other geographers situated within the participatory action research (PAR) framework (e.g. Pain 2003, 2004, Winton 2005). However, the documented use of participatory video in geographical fieldwork remains scarce and/or situated in projects at the borderline of the discipline or in an interdisciplinary setting (see Kindon 2009, Garrett 2011), and there are only very few, and rather recent, accounts of participatory video in geographies of children and young people[1] (Waite and Conn 2011).

In our paper, we reflect on a participatory video project undertaken with young people from a deprived urban neighbourhood in Slovakia. Drawing on Kindon's (2003) call to adopt video critically as a technology used simultaneously as a research methodology and as a tool for transformative empowerment, and on Parr's (2007) questioning of the relationships between various agents in the participatory video[2] process, we explore the intersubjective settings in which young participants, adult researchers and other actors pursue their individual agendas. Although we outline the context of the video project, its rationale, thematic framework, and the purpose of the video-as-a-product, the focus of the paper lies in four other themes.

The underlying theme is how participatory video is produced in an immensely complex set of relations between different actors, of which the researcher–participant relationship is only one, and often not the most significant. We reflect on our experiences, that were similar in terms of being adult facilitators of a video-making process with young people, but differed in our particular positionalities (researcher and practitioner), motivations and, ultimately, in the benefits we took from the project – even though these often intersected and some even merged. Situating the video-making process within such intersubjective settings, we simultaneously develop two additional themes. First, the paper questions how the initial motivations for taking part in a participatory video project might be different for all involved actors. It also considers the impact of such diversity in how the project is set up, and how initial motivations can differ fundamentally from the eventual benefits that actors obtain. We also focus on diversity of particular motivations and benefits even among the same groups of actors (in this case young people from the community), and discuss how this can be accounted for because of the collaborative nature of participatory video. Second, we explore what kind of relationships within and across the three types of actors in the video-making process (researcher, practitioner, young participants) are constituted, negotiated or dissolved in the process of participatory video production and dissemination. In particular, we explore distinctive ways in which participatory video affects development of these relationships; and what impact these dynamics bring for particular actors. The last theme developed in the paper is how these insights help us understand and interpret various layers and types of knowledge produced through participatory video (specifically) by young people.

The paper begins by conceptualizing participatory video in the context of both social research and community-based initiatives. We discuss the benefits and intersubjective dynamics of such collaboration, but also the problem of legitimacy and accessibility of knowledge produced by research in general. Then, we outline the context of our study, our respective positionalities, and the method employed for this paper. After this, we provide an analysis of a participatory video project undertaken with young people in Slovakia, with a focus on the complexity of relationships, motivations and benefits in the process. In conclusion, we discuss the kinds of knowledge produced through participatory video. We then seek to reconceptualize participatory video as a research/transformative tool that evolves beyond a simple duality of researcher/participant, and we outline strategies for collaboration that emphasize the notions of diversity and complementarity.

Participatory video and the dynamics of collaborative research

Participatory video can be defined simply as a production of their own video by a group of people, or, by a community. As Lunch and Lunch (2006) emphasize, 'the idea behind this is that making a video is easy and accessible, and is a great way of bringing people together to explore issues, voice concerns or simply to be creative and tell stories' (p. 10). An expanded view given by Shaw and Robertson (1997) emphasizes the notion of 'developing participants' abilities' (p. 1) through group-work, including not only technical skills but also 'engaging people in the decisions that affect them, by developing their capacities and potential and by supporting the transfer of responsibility to them to enable them to express their needs and tell their stories' (Shaw and Robertson 1997).

As such, participatory video differs from documentary filmmaking as control over, and responsibility for, the process and product stays with the participants rather than with filmmakers from outside the community. Further, the finished appearance of the video is not more important than its content and context, which are both decided by the participants (Lunch and Lunch 2006). Participatory video has a long tradition dating from the late 1960s (Lunch and Lunch 2006, Kindon 2009), but a significant benchmark in its history has been the technological changes since the late 1970s and early 1980s. Particularly since the 1990s, video equipment has become accessible also to community-based organizations and initiatives, and opened up new possibilities for interactive dissemination (Armes 1988, Shaw and Robertson 1997).

While participatory video has been traditionally associated with community-development projects, the last two decades have also been marked by increasing interest from social researchers who have incorporated video into their fieldwork practice in a way that enables the research participants to retain control over the process (Johansson 1999, Mayer 2000, Banks 2001). Experiences from research projects with children (e.g. Hadfield and Haw 2001, Haw 2005) showed that video can be used as an accessible tool for work with young people and children from disadvantaged backgrounds, while generating strong evidence-based knowledge relevant also for policy-making.

Reflections on participatory video with children and young people can bring further insights to the debates on what constitutes legitimate academic knowledge. In the discipline dominated by text (Crang 2003), the use of video provides not just a novel style of knowledge (re)presentation (Garrett 2011), but it also acts as a political tool that opens up new spaces of expression, perception and understanding for those whose active presence in academic debates is marginalized by the prevalent insistence on the use of (rigorous) language in the production of knowledge (Guidi 2003, Harris 2008, Buckingham 2009). This is especially relevant for research with children and young people, where the limits of language in encountering children's experiences have been accentuated (Horton and Kraftl 2006, Woodyer 2008). The calls for 'giving voice to children' (James and Prout 1990, James 2007), were expanded by moving not just beyond 'giving' and towards children's participation in knowledge production (Kellett 2005, van Blerk and Kesby 2008), but also beyond 'voice' and towards more multifaceted, enlivened and multisensory forms of knowledge (Hadfield and Haw 2011). Video thus begins to be seen as a tool that can not only enrich the accounts of children's lives, but can also foster children's and young people's involvement in the production of knowledge.

Participatory video is characterized by the multiple and often unpredictable ways in which it can instigate social change, or generate knowledge. Understanding video as an intrinsically participatory and transformative activity, Kindon (2009) emphasizes that 'the distinction between the use of participatory video in development or participatory video in research is less clear cut than the distinction between the emphasis on the process or the product' (p. 98). While the former distinction is often lessened if the researcher's agenda of empowerment and participative transformation is similar to the aims of the practitioners or community activists, the second distinction refers to an essential feature of participatory video that equalizes the impact that any stage of the project can possibly make. Experiences from the process of video-making, including the initial group-building and training activities, can in some cases form a firm background for future collective actions, so the eventual completion of the video is not necessarily the most significant result of the project. It does not have to be even the condition for a successful fulfilment of the transformative aims of participatory video.

Focusing on the process rather than on the product of participatory video projects opens up some additional questions. Applying the words of Davies (2000), Parr (2007) concludes that 'there are multiple orderings of film-making relating to film-making practice, production and managerial articulations of film, and through which powerful hierarchies can emerge from and between different actors' (p. 130). She, in particular distinguishes between the relationship of 'collaboration' and 'commission' on the basis of which film-making takes place. This is a crucial reminder as it highlights the multiplicity of flows in which power relations are established and (re)negotiated throughout the stages of participatory video, such as initial negotiations, field production, editing, archiving, or dissemination. However, such 'multiple orderings', while not an exclusive feature of participatory video, have other dimensions which might be crucial both for outputs of participatory video projects (including the knowledge used for research) and for the impact that the projects will have on individual actors. Several issues matter for the process of participatory video production, such as emotional dynamics of the video-making process, impact of participatory video production on relationships beyond the video-making group, or the ethical and political dimensions of the whole project (Tomaselli and Prinsloo 1990). At the centre of this debate is the question of impact on particular actors of the process, and, related to this, their motivations for being involved in participatory video production.

McDowell (2001) retells her experiences of how, in her research with young men, the 'research [was] hardly at the top of their priorities' (p. 87) and how she had to negotiate young men's motivations or willingness to be involved in her research project. She further reflects on everyday details of the research process such as representation, returning the research material to participants, or the issue of access, reconsidering what are the expectations and limits for possibilities of undertaking 'critical social research' (McDowell 2001). The lack of motivation, which she encountered with her participants, poses a risk for the aim of PAR to undertake a kind of research that will be beneficial for research participants – as Tomaselli and Prinsloo (1990) argue, fundamental questions of participation and power in participatory video are who initiates the video production and what their motivations are. From the practitioner perspective, important for our own project, there exists even an explicit requirement that the participants will gain benefit from the active relationship with the practitioner, i.e. practitioner-based involvement aims for 'striving to enhance client well-being' (McLeod 1999, p. 80). More importantly, this is what the clients also expect from practitioners (such as community workers involved in participatory video) so the conventional ethical claim of social research not to cause harm to participants (Cloke 2002, Valentine 2003) is not sufficient. Through its transformative function, participatory video should be beneficial to the participants and not just refrain from hurting them.

Our rationale for writing this paper lies in what we see as an ethical issue about knowledge production: the negotiation between different interests and motivations of different agents (including the researcher and the participants), and the effort to ensure that they will all benefit from the involvement in participatory video. We pay close attention to the intersubjective dynamics of the participatory video process because we agree that relationships in research depend on particular circumstances. They do not necessarily meet researcher's prior expectations (Alderson and Morrow 2004, Bushin 2007); so 'it is difficult for researchers to anticipate what ethics dilemmas will arise during the course of the research, [and] seeing ethics as situational and responsive is important' (Morrow and Richards 1996, p. 56). In this way, participatory video requires adopting an ethical approach that is *constantly* and iteratively reflexive to what is relevant also to other actors of this process (Skelton 2001, Elwood 2007, Manzo and Brightbill 2007, Pain 2008).

In the rest of the paper, we first set up the context of our project, and then we proceed to analyse the changing dynamics of collaboration between researcher, practitioner, and young people from a deprived community. Focusing on the issues of motivations, benefits, and interpersonal relationships, we conclude the paper with a discussion about participatory video research in relation to the notions of transformative action, efficiency of research, and the character of knowledge production.

The context of the project

The participatory video project that we discuss took place in a small residential area on the outskirts of Bratislava, the Slovak capital (Figure 1). The neighbourhood includes a lodging house funded by the Bratislava city council for families with children who cannot secure accommodation elsewhere, and a high number of council flats provided by the local district council. It is detached from the rest of the city by a railway and motorway, and it lacks basic facilities so that it is necessary to commute for health care, school, or shopping. The neighbourhood also lacks tangible facilities for children, such as playgrounds, or even benches and green areas, and no institutional services for children existed in the area until 2004, when a community centre was founded by a non-profit organization, Ulita. This provided counselling, consultancy, educational, and leisure-time activities for children and young people.

In the course of the project, the authors were both involved with a local community centre. The centre provides outreach services to children and young people from the neighbourhood, and runs a small drop-in centre with a range of structured activities for children and young people. Petra is a psychologist and community youth worker who has worked with young people in the neighbourhood from 2003. Matej is a social researcher who joined the centre for his doctoral research fieldwork that explored everyday practices and agency of children, working as a detached youth worker[3] and later also as a community worker with young people.

We discuss the first of a series of participatory video projects we ran in the neighbourhood. In this case, a small group of young people, aged 16–21, produced a video representing their account of life in the neighbourhood with the aim of presenting it at an international youth exchange in Liverpool, England (organized by the European Playwork Association, see www.go-epa.org) where they travelled with Petra and Matej in the summer of 2009. The group had been facilitated for a long time by Petra, and after her colleague stepped down from the co-facilitator position, Matej, at that time volunteering with the community centre during his doctoral project as a detached youth worker with younger children, joined the group. The group held meetings throughout the year, open to all young people from the neighbourhood, until 3 months before the international meeting. It was then closed and a structured programme was instituted focusing on group development and on preparing for the meeting. This included making a presentation about the neighbourhood and designing a workshop on the theme of the meeting ('street violence

Figure 1. Landscape of the neighbourhood.

and creative ways to tackle it') – both requirements from the organizers of the international meeting.[4] Using participatory video was an idea raised by Matej and agreed on by Petra although neither of us had any previous experience in work with video. Young people from the group consented to try video in general, although the precise nature and theme of the project were to be decided in due course. The project took place between April and July 2009, with follow-up and related activities extending until 2010. The paper is an ethnographic account of the project, written on the basis of our collective reflections. It also draws on group reflections and evaluations of the video-making processes with young people that took place regularly in the course of the project, and on interviews we undertook with the young participants after the project finished. These included questions about motivations, benefits, and relationships in the group. Details about all participants are presented in a way that secures their anonymity.

Intersubjective process of participatory video

While the paper is written jointly, we use extracts from of our e-mail communication occasionally to emphasize the differences even between our positionalities.

Initial motivations

As the facilitators of the process, our initial motivations for choosing participatory video differed, especially as a result of our positions as (primarily) a community youth worker/practitioner and as a researcher.

> My main motivation was to prepare the group for the international meeting. For this, I was looking for a meaningful activity that would develop skills such as cooperation, mutual respect, listening to each other, giving and receiving critique. Also, I expected that the activity would trigger discussions on important though sensitive themes about life in the neighbourhood, and it would transfer these discussions into a form that could be represented to different audiences, not just to participants at the international meeting. (Petra)

For Petra, as a practitioner, development of the group was the primary motivation, with a view to utilizing the constituted relationships for supporting the young people in thinking about problems, needs, or potential of their community. For Matej, the initial motivation came from the 'knowledge production' perspective – to explore young people's views of their neighbourhood in a way that would enable them to voice the 'problems' and 'questions' that they would like to raise. His actual research interest was in everyday practices of younger children from the neighbourhood, but he thought that participatory video with young people could show links to the history of the place and to changing everyday geographies of children in the area through young people's accounts of experiences of growing up in the neighbourhood.

For the young people,[5] the initial motivation was different again. They wished to attend the international meeting and were willing to take part in preparatory activities for this reason, but they had, initially, no specific interest in participatory video itself. As most of the young people said later, they welcomed the idea of trying 'something new', but their expectations about the actual nature of the activity were very vague. Also, they did not expect that participatory video would actually bring any attractive outputs and they questioned their own abilities. As one of the participants said at the end of the project:

> I thought we would produce some bullshit, anyway, so why to bother too much ... I was surprised that you were really willing to hand us a camcorder and did not think we would damage it straightaway. (17-year-old participant)

There was no clear consent about what the purpose or even the theme of the video should be. All involved actors had different motivations for taking part in the project (to 'form' the group, to generate knowledge about the neighbourhood, and to satisfy the requirements for the international exchange), but all perceived video as a tool that might help them with their agenda. There were some concerns, though: Petra thought that the focus on the video-as-a-product might dominate the group-development processes that she was mostly interested in; for Matej, using video meant a huge time investment with the risk that the participatory nature of video-making would bring results that would be irrelevant for his research interest; and the young people felt at the beginning that to produce a good video was an expectation above their capacities; so they expected the whole idea to fail.

First steps

Of the approximately 15 members of the wider group, six young people eventually formed the group that was expected to attend the international meeting.[6] The video project itself began with a residential training weekend focused on essential video skills. Of the six members of the group, only four could eventually attend.

We spent one day with introductory exercises focused on getting familiar with the camcorder, and on the second day the participants produced a short video presenting their experiences from the weekend. However, only about half of the weekend was spent with video-training activities, and the rest of time consisted of walks around the nearby lake, group-developing games, and a long evening barbecue. The first experiences with video and with video-work in the group altered the initial motivations of all actors. An important aspect of these shifts was the experience of time spent in the group and the relationships that were formed.

The biggest change in motivation happened among the young people who took part in training. While before the weekend, they understood video as an activity introduced by the group-facilitators and they agreed to take part primarily with the aim of getting a place in the group that would go to Liverpool, their first actual engagement with the video technology changed this view (Figure 2).

> You know, there's not really so much about making a video. It's pretty easy, actually. We just need help with editing, and even that does not seem so difficult, and we can do it. (17-year-old participant)

> Yes, you explained to us how to use the camcorder and we tried writing and shooting a video. We can do it on our own now, as the group. (21-year-old participant)

Three changes in the young people's motivations were important following their initial contact with video. First, they realized that making a video could be actually a pleasant time and not only shooting but also thinking about video was 'really fun'. Second, they realized that video-making, in their initial views an activity too difficult for them, was quite accessible, and they were capable of producing interesting and high-quality videos on their own, without any substantial help from others. The experience of 'success' and 'expertise' was very important as most of the group members had not experienced much appreciation before (whether in school or at home). Third, the young people reflected on the emerging relationship between themselves and the facilitators. They mentioned explicitly the participatory nature of training, where presentations of 'theoretical' skills in work with the camcorder were immediately linked to practical use of such skills, based on the participants' involvement and activity. Even more importantly, the participants expressed their feeling of trust, given and received, coming from the fact that they were

Figure 2. Video-training.

given expensive equipment and were left on their own without direct control or interventions (although a set of ethical and safety rules were agreed to by the group prior to the video-making process and they were reflected upon in the course of the project). Linking the three themes together, the group felt that they could produce an attractive video product on their own, while having fun in the process.

The initial encounters with the group in action also slightly changed our motivations as the facilitators. Even though the programme of the weekend had not focused on life in the neighbourhood explicitly, the time spent together and the emerging trust between the young people and the facilitators meant that many conversations in the group turned to young people's experiences in the neighbourhood and to the troubles, concerns, and ideas they had about the place. This meant that a body of interesting knowledge was generated in the process that illuminated contexts of children's lives in the neighbourhood, but also that the attitude of the young people to their neighbourhood was articulated and could be considered for the planning of future community-development activities. Young people's initial enthusiasm in video-making meant that they were willing to focus on work in the group, and to search for agreements and means of mutual collaboration rather than on prioritizing their own agendas. Our aims – to explore young people's experiences of the neighbourhood and to develop positive group dynamics – were thus being met already, before the actual video-making process began.

The video-making process

After the training weekend, the participants of the training, along with the facilitators, ran a workshop where they aimed to transfer relevant video-making skills to the rest of the group. A planning meeting followed where the theme for the video (an account of present life in the neighbourhood from the young people's perspective), the preliminary schedule, structure, ethical and safety rules, and organization of responsibilities, were agreed.

From this stage of the video-making process, the course of the project was in the hands of the group and we provided mostly only technical and logistical support, along with coordination of communication within the group and organization of the group meetings. However, involvement of various participants in the project was not entirely equal and power dynamics affected the video-making process, as well as relationships within the group. Two smaller groups emerged within the group – two girls who did not participate in the initial training weekend and four boys who did. One of the differences between them was the level of enthusiasm in using video.

> No, I didn't really enjoy it so much. The boys made very good shots, they are really good in that. They were better than us, maybe because of the work they did [during the training weekend]. I was happy to help them, but it was mainly their work. (16-year-old female participant)

Differences existed also within the group of boys. While some appeared to be more comfortable in creating ideas about video-shots, others preferred to provide technical support or to stand in front of the camera. Hierarchies emerged within the group and not all participants had the same voice in decisions for instance about particular shots that were recorded. Also, other young people who were not members of the group were allowed to contribute to the video fieldwork, although coordinated by the members of the group. As roles in the video-making process began to differentiate and as the gap between the boys and the girls increased, the girls were searching for a role where they felt more comfortable. Eventually, the decision was made that they would write the spoken text for the video, speak in the video, and record interviews with young people from the neighbourhood about their experiences with street violence.

Other factors affected the dynamics of the group besides individual participants' attitudes to video. Some relationships in the group, especially between individual girls and boys, were rather hostile, mainly because of some long-term personal conflicts. While the organization of the process, where everyone had a role that he or she liked and felt comfortable in, meant that everyone was involved, the actual impact of individual participants on the overall video differed. The main input came from those boys who had responsibility for field recording and activities of other group members followed their ideas and work. These 'leaders' of the process also began to appropriate the whole video as they expressed disagreement with the words or voices of the girls (but also of other boys, and eventually with use of voices in general), and requested that these be removed, arguing that it was primarily 'their' work in the field on which the video was based.

At the end of the fieldwork, there was a strong disagreement within the group about how the video should look. The main source of frustration was whether any spoken words should be actually used or if the content should be purely visual. Even though the two of us found the video produced, with spoken words, very powerful and illuminating, some of the boys in particular refused this version vigorously and required that the video be re-edited.[7] Rather than finding compromises, we (as facilitators) decided to multiply the outputs of the video process so that everyone could be involved actively in those presentations that he or she found appropriate and attractive. The girls, with some technical help from boys, made a video-document based on interviews with other young people from the neighbourhood about their experiences and ideas about street violence, which the group used in a workshop at the international meeting. The boys re-edited the main video so that it contained only a minimum of spoken words and was mostly a visual representation of the area, with some essential ideas about the life there (emphasizing negatives such as drugs, crime, pollution, and deprivation in general, but also the positive actions of some citizens). Both videos were used by the group at the international meeting. A third version of the video was completed only after the meeting, using the existing material and structure, but the narrative for the video was reworked and completed only by some members of the group who were interested in the spoken word version of the video – including both the girls and some boys.

The video was produced in the context of relationships that pre-dated the project. While the video process could not resolve the deeply rooted conflicts or power dynamics between some members of the group, it still affected the emotional and practical context of their relationships positively in two ways. First, it helped to contest the 'binary' relationship between two oppositional groups whereby, due to the hostility between some individual members, the rest of each group stood in opposition to the other group as a whole. By creating individual links through specific tasks, individual girls and boys cooperated together and found ways to improve their relationships. Second, the video served as a platform for interactions even between those individual participants who were very hostile towards each other. Rather than avoiding each other utterly, or creating conflicts within the group, they at least agreed to coordinate their activities within the overall plans of the group and were willing to search for pragmatic solutions.

An important factor of the video process was the dissemination of the video. In a series of dissemination activities in the 2 months after completion of the video – screening at the international youth exchange in Liverpool, projection of the video to the young people's families and friends, and a public community screening in the neighbourhood organized by the young people and the community centre – the young people received very positive receptions and their work was highly acclaimed both within the neighbourhood and elsewhere. At the international exchange, during a discussion that followed some themes raised by the video, especially the deprivation and negative image of the neighbourhood, the young people talked about possible positive changes in the neighbourhood and about their own potential to achieve these – a follow-up from the participatory video process that shaped the future activities of the group (see Figure 3).

Figure 3. Community screening.

Eventual benefits

Two months after the video project was finished, we undertook a group evaluation of the process followed by individual interviews between the facilitators and the participants. Subsequent to this, the group re-visited their experiences from the video-making process the following year, when they developed a plan for a series of community-development initiatives that took place in summer 2010 (including organization of a summer community cinema, sports events for younger children, and cleaning up of the neighbourhood). As a consequence, we identified the following benefits for each type of actors of the participatory video process:

For Matej, as a researcher, participatory video created a body of knowledge that he probably would not have been able to access by other means. While the young people were hesitant to talk about their experiences in the neighbourhood when asked particular questions, they demonstrated great capabilities to express their experiences and opinions if the form of expression and the scope of the problem were left to them. Participatory video enabled the young people to select the themes they wanted to present and to present them the way they wished. They produced a fascinating account of problems experienced by young people from the neighbourhood and their families, such as crime, poverty, alcohol and drug abuse, pollution, lack of opportunities, and negative images of the place. However, they also contested generalization of their neighbourhood in purely negative terms, identified positives about the neighbourhood and community, and explored solutions and opportunities for positive change. While other participatory techniques might also have been useful in this context, the visual dimension of video, as opposed to the predominantly textual scope of most conventional qualitative research methods, was highly welcome by the young people. In particular, they appreciated the dynamic nature of video, its attractiveness for (almost) any audience and its technological accessibility; for some participants, the chance to integrate the visual accounts with textual messages was also important. Moreover, even if not everything was included in the video, to be present at the video-making process meant listening to the young people articulating their experiences and participating in a mutual dialogue. The ethnography of the video-making process was as important a source of knowledge about the neighbourhood as the video-product itself. The former also helped to read and interpret the latter, where some images contained a message inaccessible to the external audience – such as when words from the video '... I grew up here with my friends ...' were accompanied by a short shot of a boy who had been taken to prison shortly before the video was finished and who was very close to some members of the group. This was a story where the background revealed much about the emotional dynamics and of the role of friendship for young people's lives in the neighbourhood.

Second, knowledge generated by participatory video was a medium that could be presented to a broader audience. Within a year of its existence, the video was shown in academic contexts (to postgraduate university students at a participatory-research class and at an academic conference) to the community, to practitioners from Slovakia and elsewhere, and to the local policy-makers in a discussion about future development of the neighbourhood. While the video shown was the same in all contexts, the accompanying story of its content and context differed. Still, the medium of presentation was praised by all audiences as accessible and revealing.

The process of creating an account of their experiences and sharing them with diverse audiences also changed the way that the young people approached Matej as a 'researcher'. At the beginning of our contact, the young people often mocked him, calling themselves 'laboratory mice' that were being investigated. While Matej attempted to challenge this view of power asymmetry in the research process, the participative experience of 'doing a research' on their own terms, of systematically[8] generating knowledge, and of seeing its impact (within the community, but also among policy-makers) changed the young people's views of what research might be and how they can themselves be involved as knowledge producers.

For Petra, as a practitioner, the video process helped to establish a group of young people who were able to reflect on their personal tensions and seek pragmatic ways of co-existence in the group. Moreover, these essentially practical skills that came from situations in the video-making process were a valuable experience for young people who learned to address and negotiate interpersonal conflicts also in their everyday life. From 'clients' – persons who were receiving services from the community centre – the young people began to become partners in activities of the centre, as they learned to deal with difference on a practical level of communication but also on a more conceptual level of better understanding others in their community. The young people also learned to be more patient and goal-oriented, and especially to manage their frustrations better. The participants themselves experienced trust from Petra and Matej as community workers who treated them in course of the participatory video project as partners. Not least, the creative engagement with problems, needs, and potential of the neighbourhood made young people not only think about possible ways of changing the place where they lived, but also to be confident about their own skills and potential, both as individuals and as a group experienced in collective – and successful – action. Additionally, the video project created material that Petra could present in her negotiations with policy-makers or media, while at the same time the existence of such material and the experience of positive reception from various audiences also encouraged young people to present their opinions in person.

Young people's account of the benefits they gained from the process drew on their initial motivations, but was expanded through reflections on their actual experiences. The initial aim was to produce a video for an international youth exchange, but as they experienced success at the initial stages of the process and as they developed their skills in video-making, their focus shifted towards the quality of the video. The participants required stricter and more intensive work within the group, and they learned to work towards deadlines, to consider details that would improve their work, and to reflect on the collective nature of the process.

> I have never before spent so much time working on just one thing ... But yes, we learned to make it good, to make it precise. And it *was* good. (17-year-old participant)

For the young people, the fact that the video process reached out to other people, including members of their own community, was important. They involved their friends or younger children within the process of field production (Figure 4) – so that some of the shots were produced collaboratively even beyond the boundaries of the group, but other community members, including adults, also saw the video afterwards and expressed their appreciation. This strengthened young people's position within the community and also their own confidence in approaching other people and being active in the community. The discussion within the group that stemmed from reflections on the video opened up the idea of young people's own involvement in community development and this led them to a series of community actions (see below) undertaken the following year. Experiences from the process of video-making helped them to self-organize and to manage the practical demands of community work. The knowledge embedded in the video-product and in the reflective process of video production then gave them a framework around which they could consider their problems, needs, and actions, and around which they built their plans for community activities in the next year.

However, the young people expressed that none of these issues would be a strong enough benefit or motivation if the video-making process had not been an enjoyable and entertaining activity itself.

> Yeah, that was great, I learned new stuff, I was not very good in all of that, but the boys learned the rest and we had great time together, working as a team, with a role for everyone ... I would not feel good

Figure 4. Field production.

in front of the camera, or even if I should make up the scene, but helping with the equipment and talking to others was my role and I enjoyed it. The video was great! (17-year-old male participant)

Yes, the boys did most of the video work, but they are good at it and they wanted to do that. But we shared our roles and I wrote the text and it was ok . . . I like the video . . . and I enjoyed my part. (16-year-old female participant)

All young participants expressed their satisfaction: with having a role they enjoyed; with the eventual results of the video production; with the fact that they managed to achieve this as a group, despite the frequent hostility between some of them; and especially with the fact that although producing such a kind of video was very demanding, they 'had fun most of time'.

In addition, the experience in the participatory video project influenced young people's subsequent involvement in the community. In the following summer, the group (along with facilitators from the community centre) organized a series of summer activities for children and other residents from the neighbourhood, made a short video about the activities, and used it for a workshop on young people's involvement in community change that they prepared and delivered at another international youth exchange. Some members of the group later expressed their interest in training younger children in the use of video and started an ongoing training in social and communication skills. Others offered their support to their younger peers who started a (non-video) project recording adult people's views about the life in the neighbourhood, offering help in reaching out to adults and negotiating the contact.

Conclusions: variety of knowledge, multitude of motivations, and integrity of benefits

Participatory video generates specific types of knowledge that can be useful for research purposes in multiple ways. The video-output is only one of the possible results and 'ethnography of the process' should be crucial for the researchers. Even such 'ethnography' can be of a collaborative nature. If the process of video-making is based on regular and frequent collective reflections that focus not only on the overall progress but also on individual feelings and motivations, then reflecting on the experiences of making a video will have a more participatory basis and can produce knowledge with an added collaborative value. Still, the video-as-a-product should not be underestimated as a research output. Its value is particularly apparent if it is also presented beyond an academic audience. Even though the academic sphere is still dominated by textual accounts, our experience was that the video produced by young people was received warmly by other community members, policy-makers, and, on a small number of occasions, also academics. Ultimately, it is often the connection between video-as-a-product and video-as-a-process that reveals deep and important aspects of participants' life experiences.

While we emphasized that young people were accepted through their work on video as 'producers of knowledge' by various adult stakeholders, it is also important to note that not only knowledge consumed by adult audiences should be counted as a valid and relevant outcome of participatory video. For one, the video in our project was screened to young people's peers from other European countries at the international meeting and the visual form, as most participants in the meeting did not speak the same language, was highly appreciated. Moreover, through our practitioner-based rationale for doing participatory video (the context of community work with young people in a deprived neighbourhood), we put a strong emphasis also on the kinds of knowledge produced by young people for themselves. These included developing their reflexivity, frustration management, and other personal and social skills, changes in personal relationships and dispositions, but also subtle emotional intelligences (Thrift 2004) about their lives and experiences – both private and intra-personal but also those reflecting on the complex and collective facets of the everyday life in the neighbourhood, and on possible alternatives and ways of change. We believe that this 'private' knowledge, produced and applied individually or within a small and relatively closed group, is not only as important as the 'outer' knowledge – produced for broader audiences or even for the public – but that it is also a necessary prerequisite for the latter as young people's self-awareness of being 'knowledgeable' helps them develop a more confident attitude towards communicating to others.[9] Importantly, we found participatory video – particularly as a tool that enabled young people to express their perspectives without depending so much on language – to be an appropriate tool for the production of both kinds of knowledge and one where the 'research' and 'transformation' elements intersect throughout most of the process.

Our experience shows that participatory video is an immensely complex activity because of the range of relationships and positionalities that various actors bring to the collaborative process. The network of actors involved in participatory video and their relationships exceed the 'facilitator/participants' duality. As we showed, even the initial roles of facilitator and participants can be driven by very different motivations and such diversity inevitably brings several implications for the practice of participatory video.

Initial motivations of individual actors to take part in a participatory video process can be – and often are – highly diverse; and they can differ significantly from the eventual benefits for the participants and facilitators. As eventual benefits can often be unexpected, it is important to reflect on the changing or expanding motivations of each actor in the participatory video process. This does not mean simply to reflect on one's own motivation and on how it changes, but rather to communicate one's motives to the rest of the group so they can be acknowledged and reflected on collectively in planning activities for further stages of the video process. Our experience illustrated that regular exchanges of feelings about the process of video-making, including what each of us expected from the project at particular moments, was crucial for sustaining our individual interest in participatory video and for balancing our expectations and plans.

The collaborative nature of participatory video is considered to be its key strength, both for the purpose of transformative actions and for the perspective of research. While motivations that individuals bring into the collective process of video-making are important for the directions that this process takes, relationships between all actors, including both facilitators and participants, are crucial for ensuring how collaborative the video-making process actually becomes. As Gallagher (2008) shows, it is not always possible to expect fully equalized power dynamics either in the participant/researcher–facilitator relationship or even among the participants themselves. Some power relations among the participants will probably pre-exist in the video-making process, and it is important that the facilitators/researchers are able to incorporate these dynamics into the video development without disturbing its collective and collaborative nature. In our story, we showed how unbalanced relationships existed among the young participants before the video project started, but also how they became even more acute and intensive because of the asymmetry of roles in the project. Mayer (2000) argues that participatory video contains high risks of privileging only some voices within the community and this argument is valid even within smaller groups involved in the project. Rather than 'flattening' the power dynamics within the group, we focused on extending the agency of the group so that each participant could find a way of individually satisfying a collectively relevant involvement in the overall process. Instead of making the dynamics of the video project homogeneous, we supported its participatory character by emphasizing its diverse nature.

Benefits of participatory video are ultimately diverse among particular actors, and many of them cannot be expected in advance if the video-making process is indeed participatory and respects individual motivations of all actors. While some benefits come from the initial motivations, some are actually results of the intersubjective character of the video-making process, particularly of the collaborative nature of the video-making activities. We see the question of benefits as a central ethical dimension of participatory video and, accordingly, the ongoing reflection and negotiation of relationships rather than relying on pre-established positionalities (see Askins 2007) as the crucial strategy for unfolding the array of possible benefits for everyone involved in participatory video. This way, participatory video can indeed become a 'collaboration' between various actors rather than a pure 'commission' for either a product or a process (Parr 2007). We also feel that participatory video is more than just a product or process of video-making and needs to be understood as an open network that contains encounters also with different kinds of audiences or with other community members that are present during, after, or even before the project. Beyond the process and product distinction, factors of dissemination and circulation of the video, and of its impact on the participants, their communities, and even on research, need to be considered carefully. While some benefits of participatory video can be common for all participants (including researchers or other facilitators), it is important to seek a kind of collaboration where individual benefits will be complementary to each other, despite the diversity of individual participants' motivations.

Acknowledgements

We very much appreciate the trust and engagement from the young people during the project. Petra would like to thank her colleagues from Ulita for their patience in coping with her initiatives. Matej wishes to thank Kye Askins and Hester Parr for their counsel on participatory video (and participatory research in general) and Fiona Smith for her ongoing support as the doctoral project supervisor. We both owe gratitude to the special issue editors and to two anonymous reviewers for their extremely constructive and supportive feedback. The paper developed during the knowledge exchange project "Community Youth Work and Social Research–Knowledge Transfer, Informing Policy and Further Opportunities for Collaboration", funded by the Economic and Social Research Council. Matej's doctoral research that involved this video project was co-funded by the Economic and Social Research Council, the Royal Geographical Society (with IBG) and the Royal Society.

Notes

1. Further examples of participatory video with children and young people can be found in other disciplines but even these are not many (e.g. Haw 2008, Moletsane *et al.* 2008). A handful of expanded accounts also exist written by community practitioners (e.g. Garthwaite 2000, Menter *et al.* 2006).

2. We should note that the project from which Parr's paper draws was not a 'fully participative video production' (Parr 2007, p. 119) and evolved from far more complex sets of expectations, initiatives, and 'commissions'. However, those Parr's arguments that we discuss here are pertinent also for the area of participatory video.
3. Detached youth work was the community centre's central method of engaging with children and young people from the neighbourhood. It is a youth work provision that takes place in young people's own territories, mainly on the street. The British Federation for Detached Youth Work gives the following description of the approach: 'As with all youth work, [detached youth work] uses the principles and practices of informal education to engage young people in constructive dialogue, within a broad agenda of personal and social development... Detached youth work, however, is distinct from all other forms of youth work as ... detached youth workers work where young people have chosen to be, whether this be streets, cafes, shopping centres etc. workers make contact with young people wherever they are. So detached youth work is often free from the constraints of centre-based youth work – where buildings are specifically set up for the purpose of youth work. This is not to say buildings won't be used; indeed they sometimes become a feature of more developed practice. But in detached youth work, contact happens on the street, and relationships are developed there too'. See http://www.detachedyouthwork.info.
4. The young people were involved in other activities during this period, but they were focused mostly on group development and had little or no connection to the video process or to 'doing research' in any broader sense. We do not discuss these activities in this paper.
5. Although we use the term 'the young people' throughout the paper, we do not want to suggest any kind of homogeneity within the group – as will be clear from our further discussion. However, the intra-group diversity was much more complex and diffused than the basic gender/ethnicity/class division could account for (see Moser 2007). We discuss these differences only to the extent we find this important for our arguments and feasible without breaching the anonymity of the participants.
6. The group was formed relatively smoothly as only these six young people expressed their interest in the international exchange and attended group meetings regularly (throughout the year before the video project began).
7. Initially, Matej did the editing, following instructions from the group after group previews of the footage. After this disagreement, some members of the group joined him for the editing, meaning they worked 6–7 h per day for a couple of days.
8. Designing the structure of the video was a difficult and painful process, especially the spoken content. In the end, we suggested a method similar to the SWOT analysis and it was approved by the group. The young people replied to the questions 'What I like in the neighbourhood, 'What I don't like in the neighbourhood', and 'What could be changed in the neighbourhood'. Reflecting collectively on individual ideas from group members generated the main body of the spoken content of the video.
9. There are links between these themes and the concept of informal education (see www.infed.org). The international exchange where the video was presented was itself funded through the European Union 'Youth in Action' programme, which promotes informal learning in young people's development. Several of the young people in our group struggled academically in mainstream education and the use of video helped them in particular to get beyond the necessity of relying on language and text. Some struggled even with reading and writing, felt awkward when this was required from them, and avoided such activities. See also Bull *et al.* (2008) for a profound discussion of connections between informal learning and participatory media and Cain (2009) for a more specific view on video in the context of community education.

References

Alderson, P. and Morrow, V., 2004. *Ethics, social research and consulting with children and young people.* Barkingside, UK: Barnardo's.

Armes, R., 1988. *On video.* London: Routledge.

Askins, K., 2007. Codes, committees and other such conundrums! *ACME: an international e-journal for critical geographies*, 6 (3), 350–359.

Banks, M., 2001. *Visual methods in social research.* London: Sage.

van Blerk, L. and Kesby, M., eds., 2008. *Doing children's geographies: methodological issues in research with young people.* London: Routledge.

Buckingham, D., 2009. 'Creative' visual methods in media research: possibilities, problems and proposals. *Media culture society*, 31 (4), 633–652.

Bull, G., et al., 2008. Connecting informal and formal learning: experiences in the age of participatory media. *Contemporary issues in technology and teacher education*, 8 (2), 100–107.
Bushin, N., 2007. Interviewing with children in their homes: putting ethical principles into practice and developing flexible techniques. *Children's geographies*, 5 (3), 235–251.
Cain, J., 2009. Understanding film and video as tools for change: applying participatory video and video advocacy in South Africa. Unpublished PhD thesis. Stellenbosch University. Available from: http://scholar.sun.ac.za/handle/10019.1/1431 [Accessed 29 February 2012].
Cloke, P., 2002. Deliver us from evil? Prospects for living ethically and acting politically in human geography. *Progress in human geography*, 26 (5), 587–604.
Crang, M., 2003. Qualitative methods: touchy, feely, look-see? *Progress in human geography*, 27 (4), 494–504.
Davies, G., 2000. Narrating the Natural History Unit: institutional orderings and spatial strategies. *Geoforum*, 31 (4), 539–551.
Elwood, S., 2007. Negotiating participatory ethics in the midst of institutional ethics. *ACME: an international e-journal for critical geographies*, 6 (3), 329–328.
Gallagher, M., 2008. 'Power is not an evil': rethinking power in participatory methods. *Children's geographies*, 6 (2), 137–150.
Garrett, B.L., 2011. Videographic geographies: using digital video for geographic research. *Progress in human geography*, 35 (4), 521–541.
Garthwaite, A., 2000. Community documentaries and participatory video. *Participatory learning and action*, 38 (June), 61–63.
Guidi, P., 2003. Guatemalan Mayan women and participatory visual media. *In*: S.A. White, ed. *Participatory video images that transform and empower*. New Delhi: Sage Publications, 252–270.
Hadfield, M. and Haw, K., 2001. 'Voice', young people and action research. *Educational action research*, 9 (3), 485–502.
Hadfield, M. and Haw, K., 2011. *Video in social science research: forms & functions*. London: Routledge.
Harris, U.S., 2008. Video for empowerment and social change. A case study with rural women in Fiji. *In*: E. Papoutsaki and U.S. Harris, eds. *South Pacific islands communications. Regional Perspective, Local Issues*. Singapore: AMIC, 186–205.
Haw, K., 2005. 'Voice' in youth activism. *In*: L.R. Sherrod, C.A. Flanagan, R. Kassimir, and A.K. Syvertsen, eds. *Youth activism: an international encyclopaedia*. Westport, CT: Greenwood Publishing, 671–676.
Haw, K., 2008. 'Voice' and video: seen, heard and listened to? *In*: P. Thomason, ed. *Doing visual research with children and young people*. London: Routledge, 192–207.
Horton, J. and Kraftl, P., 2006. What else? Some more ways of thinking about and doing children's geographies. *Children's geographies*, 4 (1), 131–143.
Hume-Cook, G., et al., 2007. Uniting people with place through participatory video: a Ngaati Hauiti journey. *In*: S. Kindon, R. Pain, and M. Kesby, eds. *Participatory action research: connecting people, participation and place*. London: Routledge, 160–169.
James, A., 2007. Giving voice to children's voices: practices and problems, pitfalls and potentials. *American anthropologist*, 109 (2), 261–272.
James, A. and Prout, J., eds., 1990. *Constructing and reconstructing childhood: contemporary issues in the sociological study of childhood*. London: Routledge.
Johansson, L., 1999. Participatory video and PRA: acknowledging the politics of empowerment, *Forests, Trees and People*, Newsletter No. 40/41, 21–23.
Kellett, M., 2005. *Developing children as researchers*. London: Paul Chapman Publishers.
Kindon, S., 2003. Participatory video in geographic research: a feminist practice of looking? *Area*, 35 (2), 142–153.
Kindon, S., 2009. Participatory video. *In*: R. Kitchin and N. Thrift, eds. *International encyclopaedia of human geography*. Amsterdam: Elsevier, vol. 8, 96–101.
Lunch, N. and Lunch, C., 2006. *Insights into participatory video: a handbook for the field*. Oxford: InsightShare.
Manzo, L.C. and Brightbill, N., 2007. Toward a participatory ethics. *In*: S. Kindon, R. Pain, and M. Kesby, eds. *Participatory action research: connecting people, participation and place*. London: Routledge, 33–40.
Mayer, V., 2000. Capturing cultural identity/creating community: a grassroots video project in San Antonio. *Texas international journal of cultural studies*, 3 (1), 57–78.
McDowell, L., 2001. 'It's that Linda again': ethical, practical and political issues involved in longitudinal research with young men. *Ethics, place and environment*, 4 (2), 87–100.

McLeod, J., 1999. *Practitioner research in counselling*. London: Sage.

Menter, H., et al., 2006. Using participatory video to develop youth leadership skills in Colombia. In: *Participatory learning and action 55: practical tools for community conservation in Southern Africa*. 107–114.

Moletsane, R., et al., 2008. Ethical issues in using participatory video in addressing gender violence in and around schools: the challenges of representation. Paper presented at the Annual Meeting of the American Educational Research Association, March 24–28, 2008, New York City. Available from: http://www.hsrc.ac.za/research/output/outputDocuments/4979_Moletsane_Ethicalissuesgenderviolence.pdf [Accessed 29 February 2012].

Morrow, V. and Richards, M., 1996. The ethics of social research with children: an overview. *Children & society*, 10 (1), 90–105.

Moser, S., 2007. Personality: a new positionality? *Area*, 40 (3), 383–392.

Pain, R., 2003. Social geography, relevance and action. *Progress in human geography*, 27 (5), 659–676.

Pain, R., 2004. Social geography: participatory research. *Progress in human geography*, 28 (5), 652–663.

Pain, R., 2008. Ethical possibilities: towards participatory ethics. *Children's geographies*, 6 (1), 95–108.

Parr, H., 2007. Collaborative film-making as process, method and text in mental health research. *Cultural geographies*, 14 (1), 114–138.

Shaw, J. and Robertson, C., 1997. *Participatory video: a practical guide to using video creatively in group development work*. London: Routledge.

Skelton, T., 2001. Girls in the Club: researching Working Class Girls' Lives. Ethics, place and environment, 4 (2), 167–173.

Thrift, N., 2004. Intensities of feeling: towards a spatial politics of affect. *Geografiska annaler B*, 86 (1), 57–78.

Tomaselli, K.G. and Prinsloo, J., 1990. Video, realism and class struggle: theoretical lacunae and the problem of power. *Continuum*, 3 (2), 140–159.

Valentine, G., 2003. In pursuit of social justice: ethics and emotions in geographies of health and disability. *Progress in human geography*, 27 (3), 375–380.

Waite, L. and Conn, C., 2011. Creating a space for young women's voices: using 'participatory video drama' in Uganda. *Gender, place and culture*, 18 (1), 115–135.

Winton, A., 2005. Using 'participatory' methods with young people in contexts of violence: reflections from Guatemala. *Bulletin of Latin American research*, 26 (4), 497–515.

Woodyer, T., 2008. The body as research tool: embodied practice and children's geographies. *Children's geographies*, 6 (4), 349–362.

Learning from young people about their lives: using participatory methods to research the impacts of AIDS in southern Africa

Nicola Ansell[a], Elsbeth Robson[a], Flora Hajdu[b] and Lorraine van Blerk[c]

[a]*Centre for Human Geography, Brunel University, London, UK;* [b]*Department of Urban and Rural Development, Swedish University of Agricultural Sciences, Uppsala, Sweden;* [c]*Geography, University of Dundee, Dundee, UK*

> Methods of participatory research have become popular among children's geographers as they are believed to enable young people to speak openly about their lives in unthreatening contexts. In this article, we reflect on our experience of using participatory methods to explore the sensitive topic of (indirect) impacts of AIDS on young people's livelihoods in Malawi and Lesotho. We examine how different methodological approaches generate varying knowledges of children's lived realities; challenges of using 'insider' and 'outsider' research assistants; the place of group-based approaches in participatory research; and ethical issues. We suggest that researchers of young people's lives should take full account of the relationship between epistemology and methodology in selecting and employing methods appropriate to particular research questions.

Introduction

Participatory research methods have been widely used in recent geographical research with young people, drawing on traditions from development studies (Chambers 1994), childhood studies (Boyden and Ennew 1997) and social geography (Pain 2004). Participatory approaches are diverse, with varied ideological underpinnings (Hickey and Mohan 2004, Kindon *et al.* 2007), and some differences between discourses prevalent in development and child research (Lund 2007). In general, however, participatory research is concerned with producing knowledge *with*, rather than *about*, those who are the subjects of the research.[1] Various methods are associated with the approach: generally, participants produce diagrams, drawings, dramas or photographs that become the focus for group discussion and collective analysis. It is not the methods themselves that make the research participatory, but rather the social relations involved in the data production and analysis, particularly with respect to where the locus of control and power lies (Gallagher 2008). These social relations involve the co-production of knowledge by a group of participants alongside 'professional' researchers. Yet, as we will demonstrate, when researching sensitive subjects, producing generalised accounts may not be the most desirable. This article reflects on our research on AIDS' impacts on young people's livelihoods and offers a conceptual contribution to the methodology of participatory research with children: that researchers of young people's lives should take full account of the relationship between epistemology and methodology in selecting and employing methods appropriate to particular research questions.

Different methods – different knowledges

This article emerges from our experience of using participatory methods to undertake research into the impacts of AIDS on young people's livelihoods in rural Malawi and Lesotho. While participatory methods proved useful in enabling young people to discuss some general aspects of their lives, they were less effective in facilitating the direct sharing of personal experiences in group contexts. Moreover, the accounts that emerged from collective participatory activities often contradicted those produced using more individualised research methods such as life history interviews (or even more individually focused 'participatory' techniques) as well as with direct observation. Diagramming methods, for instance, elicited dire stories about what happens to children when their parents die: yet, in many cases the children engaged in producing these accounts had very positive stories to tell about their own lives as orphans. Participatory research usually involves the collective production of generalised knowledges (although not exclusively so: these are also the focus of some other methodological approaches).

A number of challenges were encountered, which we explore in this article, including the different kinds of knowledges that are generated by different methodological approaches, the challenges of using insider and outsider research assistants and ethical issues. The article begins by outlining ethical and practical arguments for undertaking participatory research in general, and specifically with young people. We examine recent critiques of participatory approaches, and suggest that these emerge from a narrow epistemological perspective and that other aspects of the participatory epistemology also merit questioning. We then present some background to our research with young people in Malawi and Lesotho, our research questions and methods. We discuss the types of data produced through the research, and focus on contradictions between the knowledges produced using different methods. We examine how some participatory methods aim to build generalised accounts from individual specific knowledges, but in practice confront challenges that relate partly to the social context of the research. We conclude by suggesting that those researching young people's lives should take full account of the relationship between epistemology and methodology in selecting and employing methods appropriate to particular research questions.

Participatory research: justifications

Participatory research is favoured by many researchers, particularly those researching children and youth, for both ethical and practical reasons, or what Warshak (2003) terms 'empowerment' and 'enlightenment' rationales. Ethically, a participatory approach is considered more respectful of those whose lives are scrutinised. It entails researching *with* people, rather than extracting data *from* them and treats them not as objects but subjects in their own lives (Cahill 2004, Beazley and Ennew 2006).

From a methodological standpoint, participatory research approaches are advocated for their ability to produce 'situated, rich and layered accounts' (Pain 2004, p. 653). By involving participants in analysing the conditions of their lives, the methods are also said to be better at capturing complex non-linear inter-relationships than methods that collect descriptive data for subsequent analysis.

Participatory researchers involve participants directly in some or (ideally) all stages of research, from problem definition through data collection, analysis and dissemination to action (see Pain and Francis 2003, Kindon 2005). Innovative research techniques are often used to facilitate this, enabling participants to define and analyse problems (Kesby 2000). Non-verbal techniques frequently provide a stimulus for discussion, and may also provide data directly (O'Kane 2004).

Participatory research is said, however, to require not merely *techniques* but a participatory *process* (O'Kane 2004), whereby researchers need to 'hand over the stick' to participants, and attention shifts from individuals to the collective, with groups engaged in investigating, analysis, presentation and learning (Chambers 1994). In practice, participation levels vary greatly and the ideal ('deep participation' (Kesby 2007a)) is seldom achieved (Cleaver 2001). Although many researchers use participatory techniques without a full participatory approach (Pain and Francis 2003), the use of, for instance, visual methods without significant discussion with children of all aspects of the research process (aims, methods, level of involvement, dissemination etc.) has been criticised by Boyden and Ennew (1997) among others.

For research with children and young people, participatory techniques are understood to shift power relations, giving young people greater control over their involvement in the research. Moreover, young people's insights into their own lives are said to be most readily expressed when they are facilitated through self-directed methods (Hart 1992, Johnson et al. 1995, Boyden and Ennew 1997, Young and Barrett 2001). Different young people prefer different methods (Chawla and Kjorholt 1996, Punch 2002), so a multi-method approach enables most to contribute (Morrow 2008). Participatory research is also likely to retain children's interests, enhancing the richness of the information they provide (Punch 2002). As when working with adults, not all such research with children involves deep participation. Hart's (1997) 'ladder of participation' describes a spectrum of ways of involving young people, and although the most basic may be tokenistic and even exploitative, participation of children at the highest level is not considered universally desirable. Indeed, participants might not desire full participation; hence, it is most appropriate to work with young people on their own terms (Kindon et al. 2007).

As we explore in this article, by adopting participatory approaches to research with young people, researchers may face a number of challenges. However, first, we briefly outline some epistemological critiques of participatory research to situate our contribution in wider debates.

Participatory research: epistemological critiques

In the 1990s, participation in research was predominantly viewed as intrinsically good and rarely questioned (Cleaver 2001). By the turn of the millennium, however, it became subject to various critiques, falling into three broad categories (Pain and Francis 2003): tokenistic uses of participatory techniques, without a wider participatory approach; technical limitations such as lack of rigour, reflexivity and validity; and a broader critique of fundamental concepts (see Cooke and Kothari 2001). Although participatory research has been described as a theory of knowledge (Reason and Bradbury 2001), the key to this fundamental critique was the failure of participatory researchers to problematise knowledge production processes.

The dominant critique of participatory research emerges from a poststructuralist, constructivist perspective. Participatory methods clearly cannot directly tap objective experience or unmediated perspectives, but produce particular types of knowledge (Kesby 2000). While celebrating how participants, as meaning-making agents, engage actively in knowledge production, practitioners have attended less to the implications of the social contexts of that production (Cooke and Kothari 2001).

Two connected areas have been neglected. The first is the impact of local social inequalities and power relations (Cleaver 1999, Stokke and Mohan 2001). Participatory research emphasises local knowledge, but some individuals have the skills and authority to present their personal interests as community interests (Mosse 2001). Apparent consensus views generally conceal powerful agendas (Guijt and Kaul Shah 1998) and the multiple/conflicting knowledges held within any group or individual (Cameron and Gibson 2005). Moreover, knowledges produced not only reflect the interests of the powerful, but they can also reinforce their power (Hailey 2001, Schäfer and Yarwood 2008).

The second element of the social context is the participatory process itself, which is not a neutral means of generating knowledge. Participatory techniques may systematically facilitate certain dominant voices and subdue others. In a society where young people are not expected to speak publicly, they may effectively be silenced by a method that requires public speaking (Kapoor 2002). The effects of participatory methods on knowledge production are not coincidental. The methods are not only embedded in local contexts imbued with power, but are also products of wider power relations. The knowledge produced reflects the relationships entailed (Mosse 2001).

A number of scholars have argued that these constructivist critiques are surmountable. Participatory research is a locus of knowledge construction and potentially offers insight into knowledge processes, if engaged in with reflexive awareness of the role of the context. Kesby (2007a), for instance, attempts to reconcile participatory research and poststructuralism through attention to power relations. He argues that while arenas of participatory research are 'contrivances', 'they hold the potential to enable participants to explore the contrived nature of all social relations' (Kesby 2007b, p. 203). Participation may, in such ways, serve as a tool for social change (Cahill *et al.* 2007). Cahill (2007) considers participation as an arena for the construction of new (fluid and multiple) subjectivities. Cameron and Gibson (2005) view participatory research as a means of producing counter-stories that challenge the status quo. It is also worth noting Gallagher's (2008) observation that given the pervasiveness of power, the challenge for children's geographers is not to avoid using power in research but using it to resist domination.

Critiques of participatory methods do not always spring from a constructivist position. Other limitations of the types of knowledge produced through participatory research have been highlighted. The techniques are criticised for producing mainly linguistic representations of knowledge (diagrams, drawings and dramas are used to elicit discussion rather than providing direct insight), thus revealing little about matters that cannot be expressed verbally (Mohan 1999). Other critics have argued persuasively that through focusing on the 'local' and local knowledge, attention is shifted away from underlying socioeconomic and political forces shaping people's livelihoods (Mosse 2001, Hickey and Mohan 2004).

A perspective from which there has been little critique, but which is addressed in this article, is the relationship between participatory methods and the production of knowledge grounded in concrete experience. Participatory methods have been described as generally empiricist (Kapoor 2002); yet, the data they produce might not always relate closely to grounded realities.

Such grounded realities have seldom been at the forefront of children's geographies, which have long embraced a rather rigid epistemological stance, centred on three tenets: children are competent social actors (and therefore capable of participation in research); childhood and childhood experiences are socially constructed; and research should prioritise children's voices. The focus of much research in children's geographies has been on how young people make sense of, and are constituted by, experience/knowledge. Embedded in humanistic or poststructuralist epistemologies, both meaning-making and subject-making are arguably well suited to investigation through a participatory approach with children. However, there is a need to generate knowledge that reveals not just the meanings young people attach to experiences, but experiences themselves and how these are produced. Questions such as the one addressed through the research discussed below ('how does AIDS impact on young people's livelihoods?') seem to us both legitimate and broadly answerable. Experiences may be explored through questions such as 'what happened?' which generate answers that are shaped by individual attributes and social contexts but can also be expected to have a relation (albeit not direct and unmediated) to concrete experience. In our research, we sought to learn about children's experiences from their own perspectives, but our interest was not just in how they constructed their experiences, but how their experiences were constructed. This is a different form of knowledge from that sought by most children's geographers using participatory methods.

The research project

In 2007 and 2008, we undertook an 18-month project entitled 'Averting "New Variant Famine" in southern Africa: building food-secure rural livelihoods with AIDS-affected young people'. The research team comprised four academic researchers, including one full-time research assistant (Flora Hajdu) who undertook most of the fieldwork while resident for 2–3 months in two case study villages in Malawi and Lesotho. National steering groups, comprising potential research users, were established to advise on the research, including the appropriateness of the participatory methods selected. We also worked informally with local collaborators, and employed field assistants to help with translation.

The focus of the research was the 'new variant famine' hypothesis. This hypothesis suggests that the coincidence of extremely high HIV prevalence and recurrent food insecurity in southern Africa reflects a causal relationship: that AIDS contributes to hunger. Of the causal mechanisms proposed, several relate to the impacts of AIDS on young people. For instance, young people's livelihoods may be rendered vulnerable if household property is lost when parents die (to cover medical and funeral costs or through misappropriation by relatives); if usufruct rights to land are lost because children are considered too young to farm, or have to migrate elsewhere; or if the intergenerational transfer of knowledge and skills from adults to children is interrupted. We did not seek to 'test' the hypothesis in any formal way, but to explore the relationship between AIDS and young people's livelihoods and prospective food security.

The methods employed

A range of broadly ethnographic methods were employed in the two case study villages. The research began with community profiling workshops (Hawtin *et al.* 1994, Messer and Townsley 2003). These were intended to seek community consent for the research, build rapport, learn how people talked about the research topics and develop an understanding of the villages and their recent history. To further contextualise the information to be provided by young people in the participatory research, we undertook household profiling. The research was again introduced to every household, consent sought and basic data obtained from all households willing to participate (a very small number declined). The main data collection stage involved using participatory methods with young people. This was not 'deep' participation, inasmuch as the research questions and broad shape of the methodology had been established in advance, as is generally expected by funders, but it was conducted in a way that was broadly in line with Boyden and Ennew's (1997) prescriptions for participatory research with children. The aim was to work with young people to generate new knowledge in relation to the research questions.

The participatory research involved around thirty 10–24-year-olds in each village. They were selected on the basis of the data collected through household profiling with around half deemed to be 'AIDS-affected', usually meaning they were orphans or had experienced the chronic illness or death of an adult household member in the recent past. As in many studies of AIDS' impacts in southern Africa, chronic illness was taken as a proxy for AIDS owing to the high levels of ignorance (of diagnosis), denial and stigma associated with the disease. For most activities, the young people were divided into four groups by age and gender (girls and boys aged 10–17 and young men and women 18–24), although some opted to join a group based on their marital status rather than chronological age. In Lesotho, a (fifth) group of herd boys met at the mountainside cattle post rather than in the village. We did not distinguish between those affected and unaffected by AIDS in selecting groups or dwell on these distinctions (which were in practice somewhat blurred) during the activities, although we did focus our interest on the impacts on the participants of chronic sickness and death among household members.

The suite of participatory tools employed included drawing mental maps; daily and weekly activity charts and seasonal calendars; photography; guided transect walks; life maps; socio-spatial network and knowledge transfer diagrams; asset matrices and problem trees; emotional storyboards; and videoed drama performances. These methods were selected and developed by the research team, to ensure that they would cover all aspects of the types of knowledge required to address the research questions and would allow comparability between the two settings. However, they were often modified in light of the preferences and characteristics of a particular group and previous experience with other groups. With most techniques, participants were involved in the self- or group-directed production of a diagram, visual or dramatic output. Some, such as the dramas, were very much collective activities; others, notably the transect walks, were more individual or involved pairs of young people. In all cases, the fact that attention is not on the researcher should enable less dominant individuals to participate more comfortably, and sensitive subjects to be addressed relatively easily (Kesby 2000). Individually produced outputs were elaborated upon by their authors, and, whether individually or collectively produced, the outputs were generally then used to promote group discussion. Local interpreters translated, full notes were taken and discussions taped, transcribed and translated. In addition to these participatory activities, semi-structured interviews were conducted with key informants from the villages and local areas, as well as interviews with policy-makers and practitioners in the cities.

Several months after the original research, the team returned to the field for what had been planned as a dissemination visit. However, preliminary analysis had uncovered some of the gaps in the data that are highlighted in this article; hence, we decided to also conduct individual in-depth life history interviews to collect more specific accounts of the impacts of AIDS on young people's lives. In both villages, most resident 18–24-year-olds were interviewed. We targeted this age group due to the limited time available, and the fact that older youth could reflect over a longer life span as well as generally being more forthcoming about their lives. The interviews were conducted by Flora, Elsbeth and Nicola using outsider graduate interpreters, partly because the local interpreters tended to relate to youth in excessively hierarchical ways and were unable to translate with sufficient subtlety to capture the nuances of personal stories.

During the return visits, initial findings were also fed back to the participants through a reverse-cascade series of participatory dissemination and feedback workshops. There were three sets of workshops: with the young participants, the two communities and with policy-makers and practitioners, each group being offered the opportunity to comment on and offer their own interpretations of the findings and to feed these onward to the next workshop, principally through drama and posters. Again, graduate interpreters were employed.

The positionality of researchers is interwoven into the power relations of how we learn about young people's lives and the creation of situated knowledges. The methodological and ethical issues, challenges and advantages/disadvantages of 'insider' and 'outsider' positioning is a rich theme of reflection in the social sciences including Geography with various threads including the politics of outside researchers (Sidaway 1992), researching at home (Panini 1991, Gilbert 1994, Ite 1997), racial and gender positioning of researchers and research participants (Golde 1970, Oakley 1981, Graham 1983, Kobayashi 1994, Nast et al. 1994) and impacts of researchers' personal biographies on fieldwork and research (England 1994, Worth 2008). More recently, attention has also been focussed on personality (Moser 2008) as an aspect of positionality, positionality of outside researchers in crisis situations (Bachmann 2011) and translators (Twyman et al. 1999) whose positionality is often neglected but is also significant. We pick up this particular less-prominent theme by exploring the challenges of using both 'insider' and 'outsider' research assistants in the last section of the article.

The data produced

At the outset of the project, we envisaged that our use of participatory methods to research the sensitive topic of impacts of AIDS would generate five main forms of data: observed behaviour, and reports of facts, perceived causal relationships, and attitudes/values/aspirations, all embedded in discourse. Observed behaviour was to be noted through ethnographic observation undertaken while resident in the villages, but supplemented by information gleaned through transect walks and examination of the photographs young people took of livelihood activities. Reports of facts, causal relationships and attitudes/values/aspirations were expected to be generated through the participatory activities (although the community and household profiling would also furnish the project team with facts). Discourse would be tapped through the full range of methods, but in particular the participatory methods. In practice, however, while participatory methods proved to be a valuable means of generating knowledge in relation to perceived causal relationships and, to a large extent, people's attitudes, values and aspirations, they proved less well suited to the reporting of 'factual' information: the events and concrete circumstances that contribute to individual and collective experience.

In exploring the production of factual information, we make two further distinctions between the types of knowledges produced through different methods: between individual and collective knowledges and (closely related but not synonymous) between specific and generalised knowledges. Participatory methods are generally group methods and ultimately produce collective knowledges (as do a range of other methods such as focus groups). This is not to suggest that any knowledge is wholly individual: all knowledge is mediated and discursively produced in a social context. Yet, participatory research is frequently explicitly concerned with the production of collective knowledges. Specific knowledges are those that relate to actual conditions that pertain or have pertained, and events that have taken place, whereas generalised knowledges are those that are presented as a general truth that extends beyond a particular moment.

In undertaking our research, we were ultimately concerned with the general question: 'what happens when ... ?' (in our case when a child is affected by AIDS). In conventional empirical research, we would answer this question by exploring with a number of individuals the specific question '*what happened when ... ?*' (a close relative was sick or died). However, in participatory research, where analysis by the participants is part of the process, participants themselves address the question 'what happens when ... ?' and the relationship of that generalised answer to specific happenings or experiences may not be explicit. Moreover, it is noteworthy that collectively produced generalised knowledges are often highly normative in their construction: it is likely that the generalised answer concerns 'what should happen?' or 'what would one expect to happen?'.

The research methods fell broadly into three categories. Some were aimed purely at producing generalised (and collective) knowledges. For instance, participants developed dramas focused on problems faced by young people in the community, or on routes by which households end up hungry, followed by discussion. Similarly, during the dissemination workshops, groups produced spidergrams indicating chains of consequences following the sickness or death of a parent, and possible means of preventing these consequences. Semi-structured life history interviews, by contrast, aimed at producing specific (and individual) knowledges. The third, and most common, type of research method aimed to elicit specific knowledges and build generalised knowledges from them. In a group setting, individuals began by producing, for instance, an activity calendar representing their personal time use or a life map illustrating key events in their lives. The group then discussed general patterns emerging from these and explored patterns of difference.

Some group activities did produce specific collective knowledges, such as village profiling workshops producing historical timelines, wherein individual villagers contributed and collectively built a consensual narrative (notwithstanding the criticisms of consensus knowledges alluded to above). Similarly, groups of young people generated apparently accurate information about school fees (which include multiple elements) and livestock prices. This type of collective production of specific grounded knowledges was, however, uncommon.

Contradictions in specific accounts

Different kinds of methods resulted in different kinds of knowledges or understandings. Thus, 'factual' information generated through diverse specific data production methods was often contradictory. In a number of cases, alternative reports of facts were produced in different settings. Different people – perhaps siblings – offered conflicting accounts of the same event. Sometimes, one person would provide contradictory information on different occasions. Emily's life story included quite different details when she told it during the life maps activity from her account a year later in an individual interview. Other accounts were internally inconsistent, with, for instance, ages and years failing to tie up in many of the life history interviews, despite efforts to gain clarification (see discussion of Rex's account of his life in Ansell *et al.* 2011). Sometimes, observed behaviour conflicted with reported behaviour. On their activity calendars, for instance, some children indicated they attended school every day but were often observed in the village during school hours. Jamiya, in Malawi, said she made a livelihood by selling rice that her husband brought her from town. However, she was neither seen selling rice nor did her husband visit during the fieldwork period, and it was rumoured that he had left her. More frequently, young people said they had no source of income, or named one or two activities, but later revealed further, more lucrative livelihood activities they were engaged in. One young man in Malawi, for instance, missed a session because he was busy slaughtering a pig; butchering turned out to be an occasional source of income.

The reasons behind these alternative accounts are not always evident but may give insights into different aspects of children's lived realities. For example, it might be shameful to admit in front of friends and neighbours that one is not attending school or has no independent income. Perhaps a source of income is omitted on the assumption (however strongly denied by the researchers) that the researcher team will bring benefits to those lacking independent livelihoods. Moreover, memories are not infallible, and events will be recalled differently on different occasions. Alternatively, what is presented, even in an individual interview, might be tailored (consciously or unconsciously) to conform to an idealised or normalised version of life. That inconsistencies arise in young people's accounts is not particularly surprising or a novel finding: indeed, it is a reason for the widespread advocacy of 'triangulation' (not only in participatory research) through which multiple methods construct diverse knowledges in these areas, relating to attitudes and interpretations. Moreover, contradictions can be revealing and enhance understanding, not least by throwing up material for further investigation.

Contradictions between generalised and specific knowledges

Whatever the contradictions within specific accounts, conflicts between individual reports of personal circumstances and histories and collectively produced accounts of causal relationships were even more apparent. The following extracts are from a village-level dissemination exercise in Malawi in which groups of young people, facilitated by a field assistant, developed spidergrams (Figure 1, see also Figure 2) indicating what happened as a consequence of the sickness and subsequent death of a parent. The discussion leading to the production of the diagram was recorded.

>Assistant: How does it affect the future of youths whose parents are sick?
>Participant 1: Your future is doomed.

CHILDREN AND YOUNG PEOPLE AS KNOWLEDGE PRODUCERS

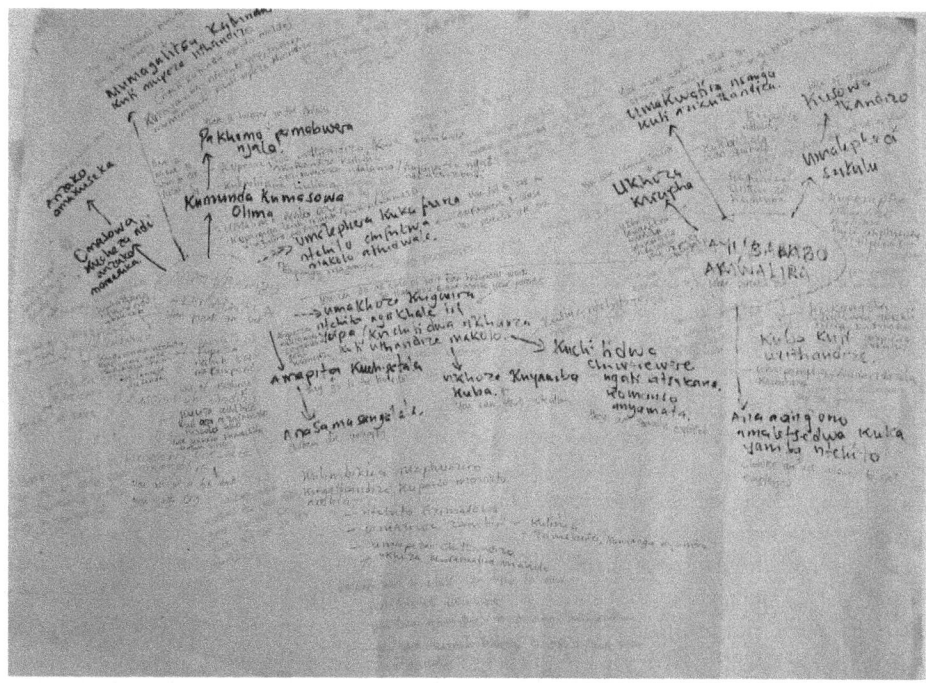

Figure 1   Spidergram produced in boys' dissemination workshop in Malawi.

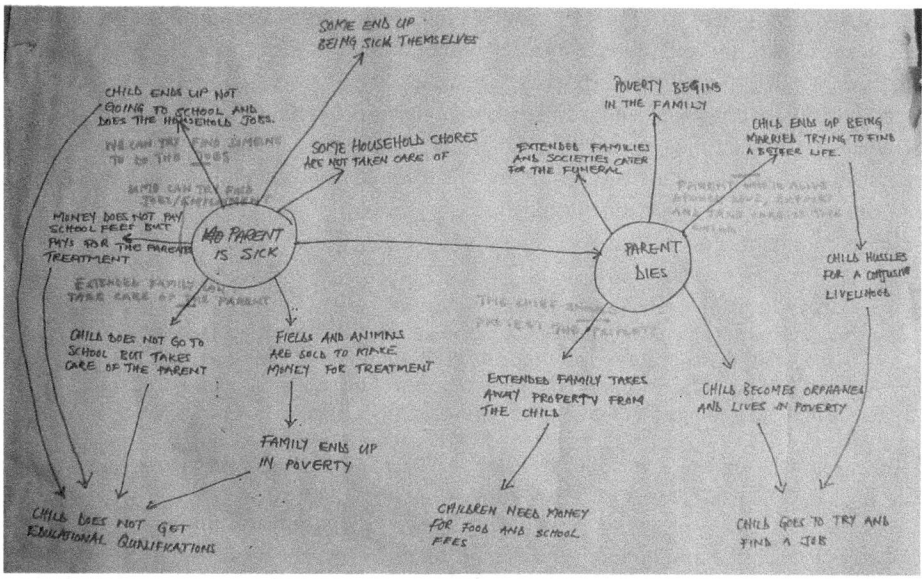

Figure 2   Spidergram produced in young people's dissemination workshop in Lesotho (translated into English).

Participant 2: Your future is doomed. If you were in school, it means your school ends there.
Participant 3: If you were doing business, your capital is used up.
  Assistant: Why is it used up?
Participant 4: It is used up because you are at home and use the money in helping your parents.

Yet this picture, which was repeated by other groups of young people and also related by adults in the community dissemination workshops, conflicted with the personal narratives obtained through other methods (both the life history interviews and more individualised elements of the participatory activities). In both villages, orphaned youth were more likely (by their own accounts) to be attending school, and those who had left had remained in school to higher levels, than those who were not orphaned. In-depth life history interviews with young people revealed no systematic differences between the livelihood activities pursued by those who were directly affected by AIDS and those who were not. That is not to deny that there were many individual stories of children who left school or were hampered in their livelihood pursuits as a consequence of parental death. However, poverty in the villages was such that most young people (irrespective of whether they were affected by AIDS) had to drop out due to lack of school fees, clothes or soap to wash their clothes, and bursaries were available to orphans, which enhanced their prospects of remaining in school.

Specific accounts from individual young people highlight many stories of exception. A considerable number of children told positive stories about their lives as orphans. Among the Malawian youth, for instance, Mary was helped by her uncle and through a bursary to complete secondary education following her father's death. David had a great deal of help from relatives to continue in school following orphanhood. Edison thought he would have to leave secondary school for lack of money, but the church choir he was active in surprised him by offering him a bursary enabling him to finish school, as his father had died. In terms of livelihoods, Emily was helped to find a job after she was orphaned. Victor and Blessings, two young orphans, were able to continue to farm their mother's field which they inherited when she died, even though they were only 12 and 10 years old, with help from their grandmother and several relatives. In the emotional storyboard activity in Lesotho, Lisebo revealed that she had been helped by the government to get shoes and clothing since she was orphaned. The one sibling-headed household in the village said their neighbours always looked out for their interests at village meetings. It is perhaps unsurprising that, when asked to produce generalised knowledge, young people do not think of specific situations of this type that might help certain individuals. It is also unsurprising that they do not think about the fact that many other poor people who still have parents have problems too. It does, however, mean that the generalised accounts generated deviate substantially from individually recalled experience.

Another group gave a very similar generalised account:

> Assistant: What is the problem when parents are sick?
> Participant: People are panicking, people are panicking a lot when thinking of their relatives/parents [who] perhaps always go to the fields. Then things at the fields won't be completed, then years ... [in] the coming year there will be hunger at home.
> Assistant: Now if things don't go well up to the extent of dying, what can happen?
> ...
> Participant: The problem is you think that since your parents have died you can start stealing, you don't worry as even if you die you can follow your mother.

The theme of orphans becoming thieves was repeated in the drama activities, along with a host of other dire consequences; yet, no mention of actual thefts in the Malawi village (of which there were several during the fieldwork period) could be related to orphans, and no young person revealed such problems when interviewed individually. The extract again reveals that collective accounts tend to produce generalised knowledges that appear to conflict with specific experiences. Peters et al. (2008, p. 34) similarly observe in Zomba District, Malawi, '[s]tereotypical opinions tend to be elicited in response to a general question about orphans, namely, that orphans tend to be neglected. When one asks about specific examples of orphans (in neighbouring families, for example), then the answers tend to be far more diverse.' In some instances, it is likely that individual accounts are censored by the interviewees. Equally, the collective accounts almost certainly drew on locally circulating narratives, as much as personal experience. Narratives about orphanhood and its consequences circulate within schools[2] and the media, as well as through the storytelling of rural communities. Chimombo's (2007) examination of the portrayal of AIDS in Malawian short stories, poetry and the arts, for instance, finds that orphans are portrayed as innocent victims, helpless, under sentence of death, vulnerable and abused. Similarly, Malawian local newspaper cuttings (2006–2008)[3] concerning orphans describe them as hopeless, needy, poor, disadvantaged, underprivileged, deprived and vulnerable to mistreatment by relatives. This negative discourse of orphanhood is reminiscent of the myth of a degraded Savannah that was co-produced between media, educational material and policy in West Africa to become taken-for-granted truth with little basis in empirical evidence (Fairhead and Leach 2003). Both 'disaster' discourses may contain elements of exaggeration in order to justify action and assistance in the form of donor and other responses. It is also possible that time plays a role. The 'myths' in circulation might not be entirely without basis, but rather describe the situation as it was in the past, before policies and programmes were introduced to assist orphans and to keep them in school. While they may be revealing of some significant 'truths', it is important to recognise that such myths do not always directly correspond to contemporary individual experiences.

Building generalised knowledges from specific knowledges

Most participatory activities began with the individual production of a drawing, diagram or other output, intended to produce personal accounts. Combining individual with collective methods to build generalised from specific knowledges proved challenging. Part of the challenge was associated with the use of drawings and diagrams, with which some participants were uncomfortable. In some cases this was because they lacked pen-handling skills, having never attended school; in others, particularly in the case of the older youth, it may have been because they did not view such activities as in keeping with their status (see Mohan 2001).[4] Even where young people took to the activity with enthusiasm, it was not always productive. In Lesotho, drawing is strongly associated with school, and activities requiring use of a pen or pencil made the atmosphere more school-like, which did not encourage discussion (although less school-like activities such as drama and photography also failed to generate much discussion). Drawing of any kind was also highly time-consuming (the life maps took the Lesotho girls nearly 2 hours to draw) and delayed the progression to discussion, which was problematic when the weather was cold and children inadequately dressed (as was often the case in Lesotho), or where the time was restricted and the young people were expected home.

Where participants were asked to draw something individually, they often did so slowly and in near silence, and when the time came for discussion they would present their own drawing, in turn, but with relatively little engagement with each other's drawings. For the activity calendar exercise in Malawi, therefore, Flora chose to begin with a group drawing, encouraging all to contribute their ideas, generating generic knowledge of daily and weekly activities, and followed this by inviting young people to produce their individual variants. With the girls, it took prompting from Flora to include school in their daily routine, although most do attend school. Reasons for non-attendance were probed, but information provided was mostly generic (a child might be sick or might need to go to the field). Alice, a 13-year-old orphan, however, used the opportunity to explain the difficulties she had encountered obtaining a transfer letter to enable her to attend the local school when she moved to live with her aunt following her grandmother's death. Nonetheless, ordering the activity in this way tended to produce generalised (albeit useful) information.

Another difficulty was that where young people were asked to represent details of their lives in the individual part of the activity, these were not always very individual or grounded in their own realities. First, because activities took place in a group setting, with young people sitting side-by-side, some were tempted to copy. On their life maps, most of the boys in Lesotho drew a hospital where they were born, a church where they went on Sundays, a grandmother whom they sometimes visited, a cattle post where they went often and a future in which they married. Among the group of young women in Malawi, all claimed to have left school early because they lacked nice clothes, although when subsequently interviewed individually, their reasons were more diverse and complex. On the 'emotional storyboard', the young people were asked to depict the happiest and saddest times in their lives, their biggest success and biggest disappointment and hopes and fears for the future. Group members often highlighted similar events. The young women in Malawi, for instance, generally claimed that their wedding day had been the happiest time; young women in Lesotho claimed marriage as the most disappointing. Such 'groupthink' has been noted by others to be a characteristic aspect of collective knowledge production which, while revealing useful understanding of the construction of discourse (Sathiparsad 2010), can be problematic when it disguises individual voices and experiences (Yuen 2004). Attempts to discourage copying sometimes backfired, with individuals presenting contrasting stories to avoid repetition, even where their experiences were actually very similar. This sometimes resulted in inconsistencies within individual accounts. In these instances, it was the group setting rather than the actual method that inhibited the production of personally grounded outputs.

Rather than copy from one another, some accounts appeared to reflect a social norm more than actual experience. Some of the Lesotho girls' activity calendars appeared aspirational or normative rather than descriptive of their daily lives. They depicted, for instance, how they sleep in at weekends and go to church every Sunday, which conflicted with their observed practices. The young married women, by contrast, depicted their daily routines as never-ending work, which might reflect the expectations of a *makoti* (new wife), as much as their actual lives.

Finally, it often proved difficult to generate group analysis of the individual drawings, diagrams and narratives. Where questions were left open, with little prompting, discussion tended to be very brief. The Lesotho boys' answer to the open question of why some boys' daily workloads appeared much greater on their activity calendars compared to others, for example, produced a single simple explanation. With prompting and probing they suggested a variety of other possible explanations, but there is a danger that discussion produced through prompting overemphasises matters closely connected with the researchers' expectations.

Moreover, there are some issues on which discussion is very difficult to generate. Describing events following the boys' life map drawing in Lesotho, Flora wrote:

> This 'discussion' consisted mainly of me asking questions, and one or another boy answering the question with a sentence or two, while turning their faces away from me and each other, shy to talk. It was very difficult to ask questions about what happens to young people when their parents pass away and they have to move. It felt like everyone thought that this was obvious – these children's lives become worse – and a strange thing to ask about. (Fieldwork notes, 12/03/08)

In this instance, some specific details of the boys' lives were gleaned through the activity, but it could hardly be described as participatory.

Social context of the research and 'insider'/'outsider' research assistants

All research methods and participatory methods in particular involve the co-construction of knowledge. Inevitably, then, the social context of that co-production affects the nature (and groundedness) of the knowledges produced. This context includes the relationships between the researcher, research assistants and the participants themselves.

The young people knew one another and, perhaps as a consequence, were reluctant to be entirely open about all their experiences. This was most apparent in the emotional storyboards activity. In Malawi, some boys declared falling off a bicycle or out of a tree to be the saddest moments in their lives, rather than referring to their parents' deaths. A girl in Lesotho drew being bitten by a snake as her saddest time. Two others in the same group also drew snakes; they hadn't been bitten but insisted that seeing a snake had been their saddest time. It seemed that these children simply did not want to share emotionally charged experiences and/or were following cultural taboos about discussing death. Such silences are not unusual among youth, particularly those who have experienced trauma. Kohli (2006), for instance, discusses young asylum seekers' use of silence in ways that are protective: as a psychological space to reflect on and make sense of their experiences; for concealing and managing hurt; and as part of the process of growing up and becoming autonomous.[5] However, while silences and absences are revealing, they do not provide accounts of concrete experience. The young people least keen to share were often those most gossiped about or ostracised. It would have been unethical to probe unwilling children further to reveal upsetting experiences and risk causing distress, embarrassment, loss of self-esteem and being re-traumatised by interview-engendered distress (Amaya-Jackson et al. 2000 cited by Alderson and Morrow 2011, p. 29). Collective methods most encourage the accounts of the more popular, thus reinforcing existing social relations as well as generating knowledge that fails to reflect the experiences of the most marginalised.

The use of, and characteristics of, research assistants also shape knowledge production. For the participatory activities, local interpreters were selected from the villages or nearby, from among the very few individuals who had completed secondary education. In Malawi these 'insider' assistants (one male, one female) were themselves quite young; in Lesotho, where there were fewer people with the requisite level of English, we selected one middle-aged woman.

For each planned activity, the local research assistants were given training. If the activity involved young people producing maps or diagrams, the assistants would learn what was expected by producing their own version and then use this with the young people as an illustration. Modelling form without influencing content proved difficult, however. The fact that most of the Malawian girls marked graveyards on their personal village maps might not signify that these were places of particular significance to themselves, but rather that the 'model' map showed a graveyard. Equally, many girls produced social network maps that included people with the same kin relations as appeared on the assistant's example (a grandmother, mother and paternal uncle). The significance attached to paternal uncles was out of line with what would be expected in matrilineal southern Malawi and conflicted with the maps produced by other groups where maternal uncles were much more important. Here, the collective activity seemingly gave undue prominence to one exception rather than simply replicating a social norm. The tendency for assistants to model was exacerbated by their lack of confidence in the younger children's capacity to understand what was expected of an activity, which led them to give examples or hints.

These local research assistants at times involved themselves more directly in knowledge production. Sometimes they appeared to 'censor' answers they did not like (for whatever reason), requiring the child to supply an alternative and denying to the researcher that a previous answer had been given. Because they were insiders, they had a complex relationship to the production of specific local knowledges. Being older and better educated than the participants, they were deferred to, and this doubtless inhibited participants from expressing certain information. At the same time, the assistants could judge the accuracy of children's testimony – which might either deter the young people from certain revelations, or equally inhibit them from telling falsehoods. At times this local knowledge and power-imbued relationship was problematic. Sometimes assistants would contradict what children said. A boy in Lesotho drew his grandparents on his social network map and said his grandmother had died in 1991. The assistant ridiculed him for forgetting that his grandmother had actually died in 2008. It was impossible to gain any real clarification of his story (perhaps it was another grandmother or other significant relative) as the boy, visibly upset, deferred to the assistant and agreed that he had been mistaken.

Given the unhelpful tendency for 'insider' researchers to act like teachers,[6] for the life history interviews and dissemination workshops we altered our strategy and appointed 'outsider' assistants, selected from a pool of graduate applicants for their experience of research and of working with young people, and on the basis of our reading of their personalities. De facto, they spoke better English and had more education/training. In the individual interviews, young people seemed very open with us, although the outsider status of the assistants meant that they could not assess with any certainty the veracity of the accounts given. However most young people interviewed had also participated in participatory activities, meaning that we had some means of gauging the reliability – mostly the interviews confirmed and deepened earlier knowledge about the young people, so it is unlikely that they were telling us 'stories'. It is difficult to gauge how far the difference between the outcomes of the interviews and the participatory research related to the characteristics and abilities of the research assistants. Moreover, the power relations underlying individual interviews, particularly with marginalised young people, are not unproblematic, as has been widely discussed elsewhere (McDowell 2001), and have a significant bearing on the knowledge production process.

The social context of the research included not only the participants, researchers and research assistants but also very often spectators. Not unusually for researchers in both Majority and Minority world contexts (Abebe 2009, Alderson and Morrow 2011, p. 38), trying to respect children's privacy was all often impossible. It proved difficult to undertake research with children in either setting without an audience, but would not have been ethically or culturally appropriate to have done the research behind closed doors even if a large enough enclosed space had been available. Undoubtedly, this (uninvited) audience had some impact on what children felt able to mention.

Conclusions

Trying to do participatory research to learn from young people in southern Africa about the sensitive issue of the impacts of AIDS on their lives raised a number of epistemological, methodological and ethical challenges which have been explored in the preceding sections of this article. Participatory research methods are principally geared to the collective construction of generalised knowledges. These knowledges at times contradict the specific personal accounts produced in other research settings. Such contradictions raise some significant epistemological questions about the status of the knowledges produced using participatory methods (or, indeed, other collective data production methods such as focus groups). A constructivist perspective on research is founded on an understanding that all research encounters produce knowledges, but none offers direct insight into empirical realities. Participatory research methods produce collective knowledges (with the researcher(s)) that have a correspondence with the empirical realities of participants' lives, but cannot be read as direct mappings of those realities. The consensus-seeking element of participatory research inevitably simplifies diversity and may lead to 'a process of controlling to produce the norm, the usual and the expected' (Kothari 2001, p. 147). This is particularly true in sensitive areas, such as research relating to AIDS, where individuals are reluctant to draw directly on their own (or others') experiences in constructing group knowledges. There is, moreover, a danger that, without direct empirical grounding in personal experience, the knowledges produced may (re)produce myths and stereotypes rather than reliable information.

This is not to dismiss the value and validity of collectively produced, generalised knowledges. While limited in their capacity to reveal lived realities, they offer insight into normative knowledges and discourses, which enhances understanding of diverse phenomena and is invaluable in informing policy and practice. It does raise important questions about the 'truths' of alternative accounts and about whether drawing on individual accounts provides 'better' data. It is worth noting that individual accounts are very diverse, and where the number of participants is small, these might not be representative of wider trends.

We believe that participatory research *can* be used to generate empirically grounded accounts. In conventional research, empirical data are gathered from individuals about their experiences, which researchers analyse to make generalisations. When undertaking group-based participatory research we encourage participants to arrive at generalisations that we usually assume are based on their personal experiences. The evidence presented in this article suggests that researchers should question such assumptions. However, rather than revert to exclusively individual methods of data collection for the construction of empirical knowledges, we propose some alternative ways to ensure that participatory methods of collective data production are more empirically grounded. One possibility is to enable participants to share their accounts with researchers initially in less public arenas. This does not preclude the use of 'participatory' techniques such as drawing and diagramming that moderate the unequal power relations of individual interviews. These methods were often productive in our own research, even where the social context of the research was not amenable to the production of grounded collective accounts. The outputs, or a summary thereof, may then be presented, anonymised, to other participants for collective analysis involving, for instance, discussion, ranking and debating (see van Blerk and Ansell 2007), thus enabling collective production of generalisable accounts in a second research stage. Another option, perhaps best suited to less sensitive subjects, is to encourage research participants to think about their evidence base or the effects of their positionality on the testimony they present. Clearly, considerable facilitation skills are required to guide participants to be critically reflexive and to present evidence. Equally, participatory ethnography might have advantages as researchers can engage with participants individually as well as collectively, producing knowledge and action with them over time and through direct involvement in their lives and communities. The ethnographic nature of such involvement can help to produce grounded accounts, although it is subject to the reflection of researchers and participants (see Blazek 2011 for an example of participatory ethnography).

In summary, while we are conscious of weaknesses in the implementation of participatory methods in our own research, and in particular the limitations imposed by working with inexperienced assistants, as well as working cross-culturally, as outsider adults, we believe there are systematic difficulties associated with using (collective) participatory methods to undertake empirical research with young people on sensitive subjects. As a consequence, our conceptual contribution to the methodological debates is to argue that when seeking to learn from young people about their lives, researchers should be more aware of the types of knowledge required, and whether participatory methods are appropriate and sufficient for generating that knowledge. It might be possible to use participatory research to produce empirically grounded accounts, even relating to sensitive subjects, but in doing so, we should consider appropriate strategies. This might involve offering opportunities for participants to share their stories initially in a more private arena or asking them to provide evidence for their assertions, rather than reifying their voices (Ansell 2009) and accepting what they say as grounded truth. Participatory methods are doubtless valuable for understanding *how* stories are produced and circulate, but there is a danger that unless accounts are interpreted critically, our research may serve to reproduce harmful myths. Finally, when seeking empirically grounded knowledge on sensitive subjects, participatory methods that require specific personal accounts to be brought into discussion in a group setting might not be the most ethical way to undertake research.

Acknowledgements

This research was funded under the joint ESRC/DFID funding scheme, contract RES-167-25-0167. We are grateful to all those who gave generously of their time in support of the project: members of the Institute of Southern African Studies, National University of Lesotho, and Department of Geography and Earth Sciences, Chancellor College, University of Malawi; the project's National Steering Groups in Malawi and Lesotho; our research assistants, translators, and transcribers; the young people and adults of Nihelo and Ha Rantelali, and all those who were interviewed for this research.

Notes

1. Our research did not focus exclusively on the present and past. As Langevang (2007) points out, combining methods enables researchers to explore young people's lives in transition, and to investigate trajectories from past events to future prospects and aspirations. In this paper, however, we focus on the production of empirically grounded conditions and events of the past and present.
2. The headteacher interviewed at the primary school in Malawi described AIDS as a key reason for children dropping out.
3. Collected and analysed by the research team.
4. It is argued that participatory methods are based in Western rationality and modes of cognition; that supposedly neutral participatory techniques such as diagramming actually rely on Western modes of seeing, understanding and representing the world and may be unfamiliar to those not educated in a Western tradition (Mohan 2001).
5. Similarly, from researching the sensitive topic of citizenship among Singaporean transmigrants Ho (2008) interrogates the many instances of silence or self-censorship.
6. Ennew *et al.* (2009, p. 2.15) emphasise very strongly that in respecting children's rights to be properly researched, researchers should avoid acting like teachers in order to minimise power inequalities between adult researchers and child participants as far as possible. This laudable ethical ideal is difficult to achieve.

References

Abebe, T., 2009. Multiple methods, complex dilemmas: negotiating socio-ethical spaces in participatory research with disadvantaged children. *Children's geographies*, 7 (4), 451–465.
Alderson, P. and Morrow, V., 2011. *The ethics of research with children and young people: a practical handbook*. London: Sage.
Amaya-Jackson, L., *et al.*, 2000. Directly questioning children and adolescents about mal-treatment. *Journal of interpersonal violence*, 15 (7), 725–759.
Ansell, N., 2009. Childhood and the politics of scale: descaling children's geographies? *Progress in human geography*, 32 (2), 190–209.
Ansell, N., van Blerk, L., Hajdu, F., and Robson, E., 2011. Spaces, times, and critical moments: a relational time – space analysis of the impacts of AIDS on rural youth in Malawi and Lesotho. *Environment and planning A*, 43 (3), 525–544.
Bachmann, V., 2011. Participating and observing: positionality and fieldwork relations during Kenya's post-election crisis. *Area*, 43 (3), 362–368.
Beazley, H. and Ennew, J., 2006. Participatory methods and approaches: tackling the two tyrannies. *In*: V. Desai and R. Potter, eds. *Doing development research*. London: Sage, 189–199.
Blazek, M., 2011. *Children's everyday practices and place: the formation of children's agency in a deprived post-socialist neighbourhood*. Unpublished PhD thesis. University of Dundee.
van Blerk, L. and Ansell, N., 2007. Participatory feedback and dissemination with and for children: reflections from research with young migrants in southern Africa. *Children's geographies*, 5 (3), 313–324.
Boyden, J. and Ennew, J., eds., 1997. *Children in focus: a manual for participatory research with children*. Stockholm: Radda Barnen.
Cahill, C., 2004. Defying gravity? Raising consciousness through collective research. *Children's geographies*, 2 (2), 273–286.
Cahill, C., 2007. The personal is political: developing new subjectivities through participatory action research. *Gender, place and culture*, 14 (3), 267–292.
Cahill, C., Sultan, F., and Pain, R., 2007. Participatory ethics: politics, practices, institutions. *ACME: an international E-journal for critical geographies*, 6 (3), 304–318.

Cameron, J. and Gibson, K., 2005. Participatory research in a poststructuralist vein. *Geoforum*, 36 (3), 315–331.
Chambers, R., 1994. Participatory rural appraisal (PRA): analysis of experience. *World development*, 22 (9), 1253–1268.
Chawla, L. and Kjorholt, A.-T., 1996. Children as special citizens. *PLA notes*, (25), 43–46.
Chimombo, S., 2007. *AIDS, artists and authors: popular responses to the epidemic 1985–2006*. Zomba: WASI Publications.
Cleaver, F., 1999. Paradoxes of participation: questioning participatory approaches to development. *Journal of international development*, 11 (4), 597–612.
Cleaver, F., 2001. Institutions, agency and the limitations of participatory approaches to development. *In*: B. Cooke and U. Kothari, eds. *Participation: the new tyranny*. London: Zed Books, 36–55.
Cooke, B. and Kothari, U., eds., 2001. *Participation: the new tyranny?* London: Zed Books.
England, K., 1994. Getting personal: reflexivity, positionality and feminist research. *Professional geographer*, 46 (1), 80–89.
Ennew, J., et al., 2009. *The right to be properly researched: how to do rights-based, scientific research with children – a set of ten manuals for field researchers*. Bangkok: Black on White.
Fairhead, J. and Leach, M., eds., 2003. *Science, power and society: environmental knowledge and policy in West Africa and the Caribbean*. Cambridge: Cambridge University Press.
Gallagher, M., 2008. 'Power is not an evil': rethinking power in participatory methods. *Children's geographies*, 6 (2), 137–150.
Gilbert, M., 1994. The politics of location: doing feminist research at 'home'. *Professional geographer*, 46 (1), 90–96.
Golde, P., ed., 1970. *Women in the field: anthropological experiences*. Chicago, IL: Aldine.
Graham, H., 1983. "Do her answers fit his questions?" Women and the survey method. *In*: G. Gamarnikow, et al., eds. *The public and the private*. London: Heinemann, 132–146.
Guijt, I. and Kaul Shah, M., eds., 1998. *The myth of community: gender issues in participatory development*. Rugby: ITDG.
Hailey, J., 2001. Beyond the formulaic: process and practice in South Asian NGOs. *In*: B. Cooke and U. Kothari, eds. *Participation: the new tyranny?* London: Zed Books, 88–101.
Hart, R., 1992. *Children's participation: from tokenism to citizenship*. Florence: UNICEF Innocenti Research Centre.
Hart, R., 1997. *Children's participation: the theory and practice of involving young citizens in community development and environmental care*. London: Earthscan.
Hawtin, M., Hughes, G., and Percy-Smith, J., 1994. *Community profiling: auditing social needs*. Buckingham: Open University Press.
Hickey, S. and Mohan, G., eds., 2004. *Participation: from tyranny to transformation? – exploring new approaches to participation in development*. London: Zed Books.
Ho, E., 2008. Embodying self-censorship: studying, writing and communicating. *Area*, 40 (4), 491–499.
Ite, U., 1997. Home, abroad, home: the challenges of postgraduate fieldwork 'at home'. *In*: E. Robson and K. Willis, eds. *Postgraduate fieldwork in developing areas: a rough guide*. 2nd ed., DARG Monograph no. 9: 75–84. London: RGS-IBG.
Johnson, V., Hill, J., and Ivan-Smith, E., 1995. *Listening to smaller voices: children in an environment of change*. Chard, Somerset: Action Aid.
Kapoor, I., 2002. The devil's in the theory: a critical assessment of Robert Chambers' work on participatory development. *Third world quarterly*, 23 (1), 101–117.
Kesby, M., 2000. Participatory diagramming: deploying qualitative methods through an action research epistemology. *Area*, 32 (4), 423–435.
Kesby, M., 2007a. Spatialising participatory approaches: the contribution of geography to a mature debate. *Environment and planning A*, 39 (12), 2813–2831.
Kesby, M., 2007b. Methodological insights on and from children's geographies. *Children's geographies*, 5 (3), 193–205.
Kindon, S., 2005. Participatory action research. *In*: I. Hay, ed. *Qualitative research methods in human geography*. Melbourne: Oxford University Press, 207–220.
Kindon, S., Pain, R. and Kesby, M., eds., 2007. *Connecting people, participation and place: participatory action research approaches and methods*. London: Routledge.
Kobayashi, A., 1994. Coloring the field: 'race' and the politics of fieldwork. *Professional geographer*, 46 (1), 73–80.

Kohli, R.K.S., 2006. The sound of silence: listening to what unaccompanied asylum-seeking children say and do not say. *British journal of social work*, 36 (5), 707–721.

Kothari, U., 2001. Participatory development: power, knowledge and social control. *In*: B. Cooke and U. Kothari, eds. *Participation: the new tyranny?* London: Zed Books, 139–152.

Langevang, T., 2007. Movements in time and space: using multiple methods in research with young people in Accra, Ghana. *Children's geographies*, 5 (3), 267–281.

Lund, R., 2007. At the interface of development studies and child research: rethinking the participating child. *Children's Geographies*, 5 (1–2), 131–148.

McDowell, L., 2001. 'It's that Linda again': ethical, practical and political issues involved in longitudinal research with young men. *Ethics, place and environment*, 4 (2), 87–100.

Messer, N. and Townsley, P., 2003. *Local institutions and livelihoods: guidelines for analysis*. Rome: FAO.

Mohan, G., 1999. Not so distant, not so strange: the personal and the political in participatory research. *Ethics, place and environment*, 2 (1), 41–54.

Mohan, G. (2001) Beyond Participation: strategies for deeper empowerment. *In*: B. Cooke and U. Kothari, eds. *Participation: the new tyranny?* London: Zed Books, 153–167.

Morrow, V., 2008. Ethical dilemmas in research with children and young people about their social environments. *Children's geographies*, 6 (1), 49–61.

Moser, S., 2008. Personality: a new positionality? *Area*, 40 (3), 383–392.

Mosse, D., 2001. 'People's knowledge', participation and patronage: operations and representations in rural development. *In*: B. Cooke and U. Kothari, eds. *Participation: the new tyranny?* London: Zed Books, 16–35.

Nast, H., *et al.*, 1994. Women in the field: critical feminist methodologies and theoretical perspectives. *Professional geographer*, 46 (1), 54–102.

Oakley, A., 1981. Interviewing women: a contradiction in terms. *In*: H. Roberts, ed. *Doing feminist research*. London: Routledge.

O'Kane, C., 2004. Responding to key challenges and ethical issues. *Children and young people as citizens: partners for social change*. Kathmandu: Save the Children Alliance.

Pain, R., 2004. Social geography: participatory research. *Progress in human geography*, 28 (5), 652–663.

Pain, R. and Francis, P., 2003. Reflections on participatory research. *Area*, 35 (1), 46–54.

Panini, M.N., ed., 1991. *From the female eye: accounts of women fieldworkers studying their own communities*. Delhi: Hindustan Publishing Corp.

Peters, P.E., Kambewa, D., and Walker, P., 2008. *The effects of increasing rates of HIV/AIDS-related illness and death on rural families in Zomba District. Malawi: a longitudinal study* [online]. RENEWAL, IFPRI. Available from: http://www.ifpri.org/sites/default/files/publications/malawizomba.pdf [Accessed 8 March 2012].

Punch, S., 2002. Research with children: the same or different from research with adults? *Childhood*, 9 (3), 321–341.

Reason, P. and Bradbury, H., eds., 2001. *The SAGE handbook of action research: participative inquiry and practice*. London: Sage.

Sathiparsad, R., 2010. Young rural males in South Africa speak on teenage pregnancy: 'It's really her problem'. *Journal of psychology in Africa*, 20 (4), 537–546.

Schäfer, N. and Yarwood, R., 2008. Involving young people as researchers: uncovering multiple power relations among youths. *Children's geographies*, 6 (2), 121–135.

Sidaway, J., 1992. In other worlds: on the politics of research by 'First World' geographers in the 'Third World'. *Area*, 24 (4), 403–408.

Stokke, K. and Mohan, G., 2001. The convergence around local civil society and the dangers of localism. *Social scientist*, 29 (11/12), 324.

Twyman, C., Morrison, J., and Sporton, D., 1999. The final fifth: autobiography, reflexivity and interpretation in cross-cultural research. *Area*, 31 (4), 313–325.

Warshak, R., 2003. Profits and pitfalls of listening to children. *Family relations*, 52 (4), 373–384.

Worth, N., 2008. The significance of the personal within disability geography. *Area*, 40 (3), 306–314.

Young, L. and Barrett, H., 2001. Adapting visual methods: action research with Kampala street children. *Area*, 33 (2), 141–152.

Yuen, F.C., 2004. 'It was fun ... I liked drawing my thoughts': using drawings as a part of the focus group process with children. *Journal of leisure research*, 36 (4), 461–482.

Critical dialogue, critical methodology: bridging the research gap to young people's participation in evaluating children's services

Liz Todd

School of Education, Communication and Language Sciences, Newcastle University, King George VI Building, Newcastle upon Tyne NE1 7RU, UK

> This paper suggests ways the academic debate on youth participation (and how this is translated into research designs) can inform the manner in which services collaborate with children and young people to evaluate their work. There is a gap between thinking from academic research and the practice of service evaluation that seems to be difficult to bridge. In order to bridge the gap, a particular kind of 'criticality' needs to be brought to the assumptions and structures of professional practice and to how research methods are used in consulting with young people. Some of the discourses of practice do not sit easily with a view of children as active agents in their own lives, a view that underlies a more meaningful collaboration with young people. A process of dialogue and reflection within services that looks questioningly at the relationship between services and young people is explored. Encouraging criticality within youth involvement in evaluation will go a long way in bringing the academic debate on young people and research to children's services. Criticality takes many forms, but this paper gives some possible areas for action. These involve research purpose, consent, method and interpretation. More youth involvement is not necessarily better. Consent is on-going rather than agreed at the start and is in the hands of young people. Rather than looking for child-friendly methods, services should think of participatory design. Conclusions from findings should go beyond taking expressed views as accessing core perspectives. Examples of participatory design that do indeed take on board the academic debate on research and young people are considered. These include the work of Investing in Children in Durham and Alison Clark's work in designing pre-school settings with children and adults such as school staff and architects.

Introduction

The gradual change in society such that children are seen as agents in their own lives rather than as those passively developing in response to upbringing and education has, over the last 20 years, brought a sea change in academic research and in the evaluation of children's services. There is also now a well-reported policy mandate to consult with children over decisions that are made about them in education and health and social services (HMSO 1989, DFE 1994, DFEE 2000, DfES 2001, Department of Health 2003, DfES 2003). However, translation of insights from the academy into the world of practice has not been easy.

Through an email, a colleague working as an educational psychologist asked my advice on how to include children's views in the evaluation that her educational psychology service (EPS) was undertaking on the quality and impact of its work with schools and families. She was looking for confirmation of the words to use in a short questionnaire for young people. I hesitated. It is not that a questionnaire was necessarily to be discouraged or that helpful comments on questionnaire design were not possible. However, what came to my mind first was the vast, on-going, complex, academic debate on youth participation and how this is translated into research designs (for example, see Gallagher 2008, Jackson and Mazzei 2009, Kellett 2010, Thomson 2008). But I struggled with my own thoughts as to how my distillation of this debate might be received by my colleague. Having worked in EPSs, I expected that practical suggestions were being asked of me rather than questions about the conceptualisation of children in children's services and the meaning of research. However, comments on the wording of the questionnaire might not bring about some of the actions that could more fully involve children in the development of services. Busy professionals in stretched services, keen to involve children, often look for practical guidance rather than for an invitation to think critically or to think more conceptually about knowledge and ideas.

This gap between research and practice also looks visible in the now substantial body of evaluative reports from different institutions looking at children's perspectives on a range of services, interventions and provisions (Kaplan *et al.* 2007, Mentoring and Befriending Foundation 2010, Moore and Dunworth 2011, to name a few examples). What appears to characterise the role of children in much of the evaluation research is the survey (by interview, focus group or questionnaire) of children's views. The manner of survey is usually seemingly uncritical, by which I mean the reporting of the findings and summaries of views, without reference to issues to do with the carrying out of the research. It is important to emphasise that this is, indeed, *seemingly* so. It may be that the critical issues that many researchers now engage with have been considered, but there is usually little indication of any critical engagement in the evaluation reports. There appears to be a gap between the critical thinking from academic research and the practice of service evaluation that is difficult to bridge. And it is these issues that this paper addresses.

This paper makes an attempt to define some aspects of this gap by naming some of the critical issues that researchers engage with that might cross the gap. I make no claim of summarising the on-going complex academic debate on youth participation and how these issues are translated into research designs as this is beyond the scope of this paper. And I do not look in detail at the reason for the gap since this would take the discussion into other issues of praxis. Instead, I am interested in how I can answer my colleague's question by referring to some aspects of the academic debate on youth participation.

I look at two areas. The first is to engage in a journey of critical reflection and cultural change in the assumptions and actions of much professional practice. This arises from research evidence that suggests dissonance between the discourse of practice and the notion of the children as 'active competent beings dealing with complex social worlds' (Christensen and Prout 2005, p. 48). The second is to engage critically with the ways this is understood in research (why involve children?) and with what is understood about consent, methods and interpretation. There is a suggestion that more child-controlled research is not necessarily better. I show that development has already been made in these areas, by drawing on two examples, the actions of 'Investing in Children' (IiC) (Cairns 2001, Williamson 2003) taking young people to participate in the development of services and the work of Clark in a collaborative approach to the design of nursery schools (Clark and Moss 2001, Clark *et al.* 2005, Clark 2010).

It may be that the last 20 years have brought about a cultural change and that the right for children to be consulted is more embedded in the cultures of professional practice and research than ever before. There is evidence that local authorities and others are continuing to support the participation of young people in their planning and delivery and to take on more critical questions and debates from research. However, the current policy context seems to be more interested in quantifying impact and measuring its cost than in bringing about and evaluating increased participation (British Educational Research Association 2011). This raises the question as to whether ground will now be lost even on past developments. Now seems a good time to take stock of some of the challenges of youth participation in evaluating services by drawing on academic debates.

Getting critical about professional practice

What is needed above all is a analytical approach to professional practice – as this informs all aspects of participatory research. The need to bring a critique to the evaluation of services comes from evidence that ideas about young people with their own agency and self-knowledge, central to much of the complex thinking in research debates, may not stand easily alongside some competing views of childhood held by professionals and embodied in assumptions of professional practices. Without reflection on, and an on-going process of evolving resolution to, these contradictions, the meaningful involvement of children in children's services does not seem to be likely. This reflection might, for example, bring Derrida's ideas about deconstruction and Foucault's understandings of modern power to professional practice.

There is evidence that children and young people, as a result of the ways that services operate, do not seem routinely to be put in a position in which they have clarity and knowledge about the roles of professionals whom they see or the possible impact of the professionals on their lives (Galloway et al. 1994, Tolley et al. 1998, Sandbaek 1999, Hobbs 2000, Grossman 2005). The notion of children as agents of their own lives, as knowledgeable about those lives, does not seem to be compatible with the prevailing deficit assumptions of the client–expert relationship when children seek professional help within children's services (Billington 2000, Todd 2000, 2007). Many discourses of practice do not always sit easily with partnership with young people. These are discourses that are embodied in referral and assessment procedures, in the manner of file-keeping and report-writing in children's services, and in the requirement of rule-following and obedience at school. There is also a literature that shows varied beliefs of professionals around childhood and, related to these, consultation that seem to be at odds with the aim of involving children in services, either in service evaluation or in decisions that involve the children themselves (Shemmings 2000, Hughes 2002, Fundudis 2003, Bailey and Wills 2010). These beliefs are not always consistent and include, for example, that children's ability to be consulted with is limited by age and other abilities (Todd 2007); that adults have a responsibility to protect children; that there is a need to contain young people and that decision-making is the role of professionals not that of children. Bragg (2010, p. 31), on the other hand, asserted that it is 'increasingly possible to challenge the idea that children are not capable of reflection or sustained engagement with complex issues' and that capability has more to do with experience in being involved in consultation and participation than with any other more essentialist and individualist considerations. These are not simple issues to debate and further discussion has been demanded even within the academy (Moore and Muller 1999, Vanderbeck 2008), calling for a recognition that accepting child agency is not unproblematic. However, there remains a need too for professionals to engage more critically in examining the relationship between their own beliefs about child agency and a wish to involve children more fully in services.

Critical engagement with the assumptions of practice is not a one-off process, but an on-going journey (Sin and Fong 2010) that involves dialogue. That this is possible and is indeed already happening is suggested by the work of Durham Council's IiC in their creative involvement of young people in public services and the work of Darlington's EPS (Hobbs et al., in press). Darlington EPS has been developing a 'narrative' approach to practice. Over a period of five years, the Darlington service has gradually changed the ways it practises.

Another version of how to bring about a critical change in practices with children is exemplified in the ways that IiC works with professionals. Williamson and Cairns (2005, 2006) reported on two conferences and a workshop with children's service professionals from all over the country, including social workers, children's rights officers, family support workers, counsellors and service managers. The conferences involved a series of tasks designed to help people reflect on principles of effective practice in projects at work where they had attempted to promote the participation of young people. Practitioners were invited to bring along accounts of their own participation experiences. They discussed a presentation of observations from IiC's work that included some thought-provoking claims such as the view that challenging the political marginalisation of children requires that we find ways of supporting young people to assert their right to be seen as citizens. This was designed to 'hold up the mirror' to participants to help reflect on their work. The conference organisers promoted the development of a community of practice, fostering dialogue with an atmosphere of active listening in an open, non-judgemental way to what other people were saying.

The main issues that were discussed in the conference that needed to be addressed were relationships and communication. A need was identified to reflect on 'features of the structure of power within society and of the ways in which those structures are made legitimate through a policy discourse that both marginalises and often demonises young people' (Williamson and Cairns 2005, p. 14). They quoted one participant (p. 8):

> 'Adults' one colleague commented, 'often declare how disappointed they are with children, but children cannot be disappointed with adults!' Adults have to work hard to create the conditions of communication that will promote non-distorted listening. One colleague noted that 'Attitudes towards young people are stereotypically negative – "ungrateful", "difficult", "problematic"'.

This is a conceptualisation of practice change that involves political as much as technical and interactional change. They concluded (Williamson and Cairns 2005, p. 2), after the first two conferences, that

> current policy objectives in several fields of children's services and education to promote citizenship and participation in shaping service delivery are severely limited. Principles are one thing; practice is another. Too often the aim of engaging young people is vitiated by existing structures of professional power and cultural attitudes that devalue the opinions and skills of young people. Without managed cultural change young people will continue to feel marginalised in this society. The good news is that attitudes can be changed but not without changes in the institutions that govern the lives of young people.

Bailey and Wills (2010), talking about theoretical and practical issues to be considered in trying to build a culture of children and young people's involvement in decision-making in children's services, came to similar conclusions that 'Children and young people's involvement in decision-making is a multi-dimensional challenge that requires organisational change' (p. 80). That such a change is possible is exemplified in the work of IiC and the work of their many partner services and the service delivery of Darlington EPS (Hobbs et al., in press). What all have in common is a critical approach to practice. For example, young people from IiC have worked with Durham children's services to develop rules of practice that take young people's views into consideration at every stage of service delivery (Investing in Children 2011). Darlington EPS has evolved over time, through critical dialogue within the team of practitioners, a narrative approach to practice (White 2007). This approach views problems as separate from the person, provides a challenge to prevailing socio-cultural narratives and offers alternative richer understandings of behaviours often characterised as problematic. This has led the service to take on a number of ways of working professionally that offer more potential for children's participation.

Getting critical about research

Experienced practitioners, talking to me about their ideas for post-graduate research, sometimes appear surprised when I ask them why they want to ask children and young people their views. Hart introduced the 'Ladder of Participation', a model of participation through eight levels, starting from manipulation and non-participation and moving up towards equal participation of adults and children (Hart 1992, Shier 2001). This has helped many researchers and practitioners to move to greater participation. However, it can encourage the view that involving children more is always the way to go. The need is rather for a critical research-based approach to evaluations that involve children – and this does not mean that being further up the participation ladder, away from adult control, is necessary and is always better. The logical and simple appeal of Hart's deservedly popular visual and linear conceptualisation ignores the political complexities that shape the production and reception of the child in research. A more nuanced approach that looks critically at purpose, consent, method and interpretation in research with children is needed.

Purpose

So, what kind of understandings inform how we think about young people being involved in evaluations in services? Young people are assumed to be lacking in power and less likely to be in a position, for example, to influence decision-making in areas that concern them. This, therefore, leads to efforts to bring their perspectives into the picture (Hacking *et al.* 2007a, 2007b, Kaplan *et al.* 2007, Marquez-Zenkov *et al.* 2007, Nystrom 2007). However, one of the things that child voice research has lacked is a critical rationale. The orthodoxy within some areas of professional practice that under-represented voices need to be heard has, perhaps, had the effect of silencing critique.

Without this critical rationale, there is a risk of disempowering young people even further. One way that this can happen is when children's heterogeneity is ignored and when children are differentially included in research by treating them as a homogenous group. Even in the genuine efforts of professionals to involve them, children risk being co-opted onto professional viewpoints (Galloway *et al.* 1994, Todd 2007). McDowell *et al.* (2005) and Schäfer and Yarwood (2008) have warned against possible disempowering potentials of research and also highlighted the multiple-power relations among young people which affect the research process. The context of a rising orthodoxy of child consultation means that there are many different and indeed competing interests that are met in 'voice' research. Without a culture of criticality about purpose, method or meaning, there is a risk of tokenism when schools, for example, consult children routinely as part of various disciplinary frameworks (Ofsted 2003). Whitty and Wisby (2008) indeed asked to what extent are pupils allowed a 'voice' other than to legitimate local, government or school policy. Such tokenism can also be seen in many school councils (Alderson 2000, Whitty and Wisby 2008) and the consultation of young people by some agencies about services and local and national decision-makers about policy. However, some government departments have shown willingness to counter tokenism (DfES 2004). There seems to be a lack of self-awareness, even among researchers and practitioners who seek social justice for children, that those who practise youth consultation and involvement have much to gain themselves in terms of social and other forms of capital.

Much consultation of young people is motivated by a market-based rationale, with the child cast in the role of a consumer. This brings potential hazards for young people. The neo-liberalisation of social, health and educational institutions has meant that researchers risk contributing to a co-option of children further into 'consumerism' and into 'responsibilisation', making children feel responsible for the achievement of social policy such as good schools over which they can have no real control (Whitty and Wisby 2008). As Bragg (2010) said, 'Even participatory measures that are benign in intent may contribute to this neo-liberal agenda by obscuring the structural factors behind inequalities, in favour of individual responsibility-taking in a "power-loaded game" (Triantafillou and Nielsen, 2001)' (p. 22). However, there are indeed resistances. I am reminded of a young person who, in effect, took a stand against such 'responsibilisation', when, having given his views on Ofsted school inspections, refused an invitation to be part of a project looking at ways to change the inspections. 'No', he said to the educationalist, 'that's your job, what you are paid to do!'

In academic literature, however, the rationale for involving children has long moved beyond the need to involve children by virtue of being one group of the 'neglected others' whose geographies needed to be accounted for (Philo 1993). Murdoch and Pratt (1993) similarly reflected critically arguing that it was not sufficient to focus on 'neglected others', but to understand the power structures which caused social inclusion and exclusion.

This led to the call for a social science of childhood (James and Prout 1990, Brannen and O'Brien 1995) which highlighted young people's agency and provided a theoretical framework which allowed for the development of a more critical rationale for the potential, opportunities and needs to involve young people in research. The intentions and actions of children in the 'reproduction and re-interpretation' of society became the focus of study (Christensen and Prout 2005, p. 50). It followed that children could be understood as active – rather than passive – in their interaction with research situations and with meaning-making about the research process itself. When the questions 'why consult children?' and 'how are voices produced?' (James 2007) are asked, we are led to consider both the position of and the construction of the child in society and also the nature and purpose of research. One cannot be considered without the other.

Getting back to the experienced practitioner to whom I asked 'why do you want to consult with children?', I am suggesting the need to think through the purpose, and what might be an appropriate methodology, and how might the findings be interpreted. If the purpose is to involve children in decision-making about their own future, this might involve more than a questionnaire on their likes and dislikes about school. If the purpose is to evolve better children's services, just asking questions about what is good and bad about a service may not be sufficient to put children in a position in which they are able to comment on service delivery. If the purpose is to consult on the experience of teaching and learning of children who have particular special needs and who are included in mainstream classes, asking children to do a paper or verbal task on, say, what they think makes a good teacher is unlikely in itself to constitute consultation. We are, therefore, led to reflect on who is being asked, about what, how, to serve whose interests and for what purpose (James 2007, Bragg 2010). This leads us to the methodology and methods by which children are involved in research.

Consent

Consent seems to be one area in which there are clear lessons from academic research for the practice arena. There is a large research literature about medical consent, exploring questions about children's consent to take part in research and medical procedures. Consent has a complex literature that is beyond the scope of this paper to discuss fully (Balen *et al.* 2006, Leitch *et al.* 2007, Flicker and Guta 2008). One of the most comprehensive discussions of ways of thinking about consent is Alderson and Goodey's (1998) analysis of theories that relate as much to children's consent around educational research as to that within medical decisions. They suggested positivist frameworks encourage accountability and the clear provision of information to enable consent decisions to be made. Social constructionist ideas help an understanding that the process of thinking about whether to give consent is complex, constrained by a range of influences. Post-modernism shows the hidden pitfalls of the morphing of consent into choice, particularly in today's neo-liberal context.

The kind of consent debated in academic research aims to put children in a position in which they are able to choose to take part or to not take part in research. What makes this problematic in schools and children's services is the highly constrained nature of consent in many professional relationships with children which is bound up with legal institutional requirements and with discourses of the child. It is not easy to negotiate consent when, for example, children are required to attend school and obey the rules in lessons, and when they may feel that they are expected to comply with the assessment practices of children's service practitioners. Ideas of the child as dependent and vulnerable, developing into 'adult', has in the past deprived children of the right to give consent. Thus, teachers and parents have given consent on behalf of children for them to take part in research, particularly in school settings, thereby placing children without agency.

Decisions regarding consent taken to protect children may run counter to their rights. Flicker and Guta (2008) and Balen *et al.* (2006) argued that consent through parents is inappropriate The role of ethical committees (i.e. of universities and health boards) in consent is an area that deserves further research. What is in the institution's interests may not always be in the interests of children, and this has sometimes led to adult refusal to allow research that may be welcomed by young people (Campbell 2008).

Consent needs to be considered even before the point where young people are asked to give agreement to be involved in research. For example, there are ethical issues in researching young people as presumed members of particular groups or populations (i.e. children living in poverty, children who suffer particular illnesses and 'looked' after children'), when it cannot be known prior to approaching them whether they are in agreement with being approached. A young person may not wish to be known or described as, for example, 'living in poverty', and he or she might not think in terms of such a label. Being invited to a meeting even to talk about this in a semi-public setting such as a school could have the child being seen by others in terms of labels with which he or she does not hold.

It is not easy for ethical researchers to get it right (Robson *et al.* 2009). Even where children are involved in careful consent considerations, they can find that half way through the research, their agreement to participate wanes as they do not like having to miss school sessions as a result of participation in the research. What has evolved in a number of disciplines is the notion of consent as on-going and reviewed over time, rather than just as a one-off agreement (Cocks 2006, Morrow 2008). This approach is similar to that of Clark (2010), who sees research as an 'ethical relationship between adults and children' rather than as a set of methods (p. 15).

Methods

Methodological critique in children's services – and also in academic research – needs to go beyond a more pragmatic search for what is 'accessible' or 'appropriate' for children and the assumptions made about what these ('accessibility' or 'appropriateness') signify. Others have argued for a participatory design (Cooke and Kothari 2001, Punch 2002, Kesby 2007, Clark 2010) rather than for participatory methods. The adoption of methods as seemingly more 'child friendly' misses the complexities of the research situation, as a contested and constructed site. For example, Cook and Hess (2007) discussed different methods that are designed to be 'fun' such as drawings, photographs, participatory rural appraisal, diaries and worksheets. Although photography was indeed perceived by children as fun, Cook and Hess suggested that researchers and children seemed to perceive the research aims in quite different ways when using them. A question is raised as to whether their use in research would lead to the purpose expected by the researchers. Cook and Hess also raised ethical issues in giving new technology to children who have not used cameras before if such use encourages expectations that may not be fulfilled, that is, that the cameras can be used after the research and that children think that they can become photographers.

Other researchers have tried to find ways around the dilemmas of appropriateness by involving young people in the development of the methods (Thomson and Gunter 2006). Thomson and Gunter worked with students to investigate the way one school went about improving student learning through putting in place choice pathways through the curriculum. To do this, they evolved what they called 'standpoint tools' recognising this as standpoint research (students as insiders to certain kinds of knowledge). This involved the development by students of photos of people and belongings to use with other students in photo-elicitation.

Reay (2006) made the case for researching the perspectives of pupils without their direct involvement in the research process. She explored the ways in which pupil consultation can uncover aspects of both pupil peer group cultures and classroom dynamics that work against aspects of fairness, collegiality and community. Her intention was that otherwise hidden pupil concerns are raised and are 'issues that pupils do not feel comfortable about talking directly with their teachers and yet are happy for the researcher to raise on their behalf' (p. 171).

Clark's methodology, developed over the last 10 years, is a good example of participatory design that involves very young children (Clark and Moss 2001, Clark et al. 2005, Clark 2010). In one project, children and adults were actively involved in the design, development and review of early childhood centres and schools (Clark 2010). Clark used an approach that is 'multi-methods, reflexive, adaptable, focused on children's lived experiences and embedded into practice (i.e. such as teaching)' (Moss 2010, p. 11). The aim was to involve young children's views and experiences such that they informed the planning, design and development of early years' provision. The process involved gathering perspectives, facilitating exchange and influencing practice. The assumptions about children that informed this research and development were as follows:

- young children as experts in their own lives,
- young children as skilful communicators,
- young children as active participants and
- young children as meaning-makers, researchers and explorers.

The 'mosaic' approach, developed with other researchers (Clark and Moss 2001), was used. This involves the judicious and careful putting together of data from different sources. Among the methods that form part of a mosaic approach are children's interviews, photo books, model-making, slide shows, observations, tours, map-making and parents' views. There is an emphasis on knowledge construction rather than on knowledge extraction. This creates certain possibilities. Researchers are enabled to 'see' differently and the status of knowledge created by participants is raised. Communication at different stages of the research process is facilitated. Research outcomes, the data, are then 'subject to review, reflection, discussion and an interpretation by children and adults in a process of participant meaning-making' (Moss 2010, p. 11). Clarke (personal communication, April 5, 2011) gave an illuminating picture of children's story boards of their pre-school setting being pondered over by the architects. To me it seemed that the story boards were more useful to the architects than, say, certain data that might have been created by adult researchers, as the boards are artefacts used commonly by architects and ones that speak of their ways of thinking and working. I could see how knowledge creation could happen.

By contrast, IiC, funded by Durham County Council, has developed practices, strategies and policies designed to engage older children actively in consultation and decision-making processes in matters affecting their lives. One model of working developed by IiC is to undertake an 'agenda day' which (typically) 10–25 diverse young people aged 13–18 years meet in an adult-free environment to express their views, facilitated by other young people who have met before to plan the day. The facilitators plan and decide which questions are to be asked of the invited young people that would give them the best opportunity to express their views. The outcomes of agenda days have contributed to the development of many different organisations including schools, health services, sports facilities, the delivery of social support services and many more. IiC is constantly evolving the methods that they use and young people are fully included in this development.

Where researchers are also the practitioners, the notion of the research situation as a joint construction of the young person and the practitioner suggests the need for reflexivity and for critical reflection. One aspect of this reflection is about the role of, for example, social workers, psychologists or teachers and their effects on the research. While there is an extensive literature on researcher reflexivity (Burgess 1984, Guillemin and Gillam 2004), the awareness of the impact and import of researcher identity and perspectives on the research context and critical recognition of such issues seem to be more visible in the academic literature rather than in the applied one.

However, young people's critique of research methods seems to implicitly recognise the situation of research and the researcher as constructive of the response. For young people, liking the questioner was important, as was an informal conversational style. It was important that there be no right or wrong answers. The language used by those consulting with children was seen to be very important – that people should avoid jargon and should use vocabulary that young people themselves would use:

> parents and professionals wishing to discuss mental health with young people should either use alternative vocabulary based on young people's own, or be very careful to identify what they mean. (Armstrong *et al.* 2000, p. 69)

Some young people were very much aware that it would be difficult if they knew the person asking them for their views. If the young person knew and liked the staff questioner, this could, it was felt, inhibit negative comments for fear of giving offence or getting staff into some sort of trouble. Young people advised staff to consider a questioner from another department or from outside the organisation (Lightfoot and Sloper 2003). This has particular implications for the evaluation of school activities and services although such issues seem to be rarely acknowledged in published evaluations.

Researching with children involves 'queer(ing) the relationship between the researcher and the researched' (McClelland and Fine 2008) in order to 'rupture ... the ordinary' (Leitch *et al.* 2007). This might have us adult researchers adopting Barkty's (2008) kind of reflexivity that involves 'world travelling' ('...by travelling to their worlds we can understand what it is to be them and what it is to be ourselves in their eyes') and 'playfulness', 'a lack of abandonment to a particular construction of oneself, others and one's relation to them.' (p. 78).

Interpretation

The interpretation of findings is not a simple matter. What is to be made of young people's responses within an evaluation and what does academic debate on participation bring to this question? An example from my own experience while supporting others in their local authority research may be illuminating. An evaluation of the ways that inclusion for young people with disabilities was put into practice in a secondary school was carried out by a colleague using interviews with a group of young people who spent most of their break and lunch times in a room set aside for them. All the children had been deemed to have special educational needs such that additional staffing was given to them to assist in their education. As part of the research, the young people remarked on the helpfulness of all the teachers and that there were so few demands on them as children. They were always accompanied to lunch and escorted back to their base room and never felt worried.

Limited though this example is in detail, a possible conclusion is that the children were indeed successfully included into school. A negative interpretation might be that an example of social exclusion has been expressed, as young people are limited to particular spaces in the school through a level of surveillance in their lives that might inhibit other social relationships.

What seems to be the issue here is the meaning given to the views of the children, as to whether conclusions stop with the reported views or whether the views denote something about the social and structural positioning of the young people in the school. One can speculate that a long-term ethnographic research approach, being alongside the young people, might have thrown up a number of concerns experienced by these young people about life at school. This may have shown a more complex picture of inclusions and exclusions, related to both school practices and the different identity positions of young people. Allen (1999) noticed that children with disabilities in schools liked to choose when to take on different aspects of their disabled identities.

There are merits in moving away from seeing the purpose of seeking children's views solely in terms of neglected perspectives. In addition, we can look at what children's views tell us about power structures and the causes of social inclusion and exclusion. Of course, there may be a concern that conclusions from data may lead further away from practice into the realms of the theoretical framework. However, if indeed (and of course we can only speculate from such a brief example) these young people were experiencing social exclusion, rather than just caring inclusion, a critical approach to this evaluation might indeed draw the practitioner to a more appropriate action. This brings to mind that 'There is nothing so practical as a good theory' (Lewin 1951).

The absence of a critique seems to lead to the uncritical treatment of voice. It can also lead to a disregard for issues of gender, race and class in children's relationships with schools, as we can see in the relationship between parents and schools (David 1994, Reay 2004, Crozier and Davies 2007). What this critique might look like is a change in thinking, away from seeing views as somehow what young people really think 'as if we can simply access their authentic core being' (Bragg 2010, p. 31) to understand views as produced by/within dominant discourses. It is likely that 'simply having the right to speak and research did not mean that what was said by students was somehow any more "pure" or "authentic" than any other voices (including our own)' (Thomson and Gunter 2006, p. 852).

Thomson and Gunter (2007, p. 330) quoting Alenen (2005, p. 43), suggested that '"the truth regime in which children are implicated" produces particular knowledges and experiences about social relations which are "normally not articulated, and remain hidden, implicit, unacknowledged"'. This is a social production of knowledges and is not the same as the linear addition of missing views that are regarded in more or less essentialist ways and that are the assumed responses of the not-yet-adult. Moore and Muller (1999) discussed concerns that 'voice' research assumes the reduction of knowledge to the single plane of experience. Research can yield many different understandings of knowledge, given that 'Our speech is filled with other's words, varying degrees of otherness or varying degrees of "ourown-ness"' (Bakhtin 1986a, 1986b).

Concluding comments

Without critique, there is a danger that research that simply asks children their views lacks rigour, puts ethics and purpose at risk, and courts tokenism. I started with practice – but have argued for the need for a dialogic process of critique, looking for a gradual change in the culture of professional practice. Where the discourses and assumptions of practice are at odds with a view of child as a knowledgable agent in the world, this does not provide a helpful basis from which to start to develop meaningful collaboration. There is a need to take a step into unknown territory and ask questions: 'If we were all more lost we would be better off' (Caputo 1997, p. 57).

This paper has argued against simply increasing child-controlled research and instead sought to develop a critical approach to research. A lack of criticality risks more marketisation, more responsibilisation and the further disempowerment of young people. Some very particular aspects of a critical approach to research were considered, selecting those most likely to be helpful to services. Rather than consent as a single event, think of children as giving initial consent, but then continue to think of consent throughout the research as an on-going ethical process. Instead of consulting children due to their voice having been neglected in the past, a more strategic purpose that seeks to contribute to conceptual debates can be more helpful. Participatory design is more likely to bring about participation than looking for the so-called child-friendly data collection methods. We move beyond children's 'views' or 'voice' as synonymous to 'knowledge' to be aware of the multi-voiced nature of all interactions. There is no simple equating of the greater involvement of children in research with progress and social justice.

References

Alderson, P., 2000. Children's rights and school councils. *Children and society*, 14 (2), 21–34.
Alderson, P. and Goodey, C., 1998. Theories in health care and research: theories of consent. *British medical journal*, 7 (317), 1313–1315.
Allen, J., 1999. *Actively seeking inclusion. Pupils with special needs in mainstream schools*. London: Falmer Press.
Armstrong, C., Hill, M., and Secker, J., 2000. Young people's perceptions of mental health. *Children and society*, 14 (1), 60–72.
Bailey, M. and Wills, W., 2010. Undertaking and building a culture of child/youth involvement in a context of children's trust partnerships. *Children's geographies*, 8 (1), 79–84.
Balen, R., et al., 2006. Involving children in health and social research: 'human becomings' or 'active beings'? *Childhood*, 13 (1), 29–48.
Bakhtin, M.M., 1986a. *The dialectical imagination: four essays* (M. Holquist, ed.). Austin, TX: University of Texas Press.
Bakhtin, M.M., 1986b. *Speech genres and other late essays* (trans. by Vern W. McGee). Austin, TX: University of Texas Press.
Barkty, S., 2008. On psychological oppression. *In*: A. Bailey and C. Cuomo, eds. *The feminist philosophy reader*. New York: McGraw-Hill.
Billington, T., 2000. *Separating, losing and excluding children*. London: Routledge Falmer.
Bragg, S., 2010. *Consulting young people: a literature review*. 2nd ed. Newcastle upon Tyne: Creativity, Culture and Education.
Brannen, J. and O'Brien, M., 1995. Childhood and the sociological gaze: paradigms and paradoxes. *Sociology*, 29 (4), 729–737.
British Educational Research Association, 2011. *Research intelligence. The current state of educational research. Issue 115*. London: British Educational Research Association.
Burgess, R.G., 1984. *In the field. An introduction to field research*. London: Routledge.
Cairns, L., 2001. Investing in children: learning how to promote the rights of all children. *Children and society*, 15, 347–360.
Campbell, A., 2008. For their own good: recruiting children for research. *Childhood*, 15 (1), 30–49.
Caputo, J.D., ed., 1997. *Deconstruction in a nutshell: a conversation with Jaques Derrida*. New York: Fordham University Press.
Christensen, P. and Prout, A., 2005. Anthropological and sociological perspectives on the study of children. *In*: S. Greene and D. Hogan, eds. *Researching children's experience: approaches and methods*. London: Sage.
Clark, A., 2010. *Transforming children's spaces. Children's and adults' participation in designing learning environments*. London: Routledge.

Clark, A. and Moss, P., 2001. *Listening to young children: the mosaic approach.* London: National Children's Bureau and Joseph Rowntree Foundation.
Clark, A., Kjorholt, A.T. and Moss, P., eds., 2005. *Beyond listening: children's perspectives on early childhood services.* Bristol: The Policy Press.
Cocks, A., 2006. The ethical maze: finding an inclusive path towards gaining children's agreement to research participation. *Childhood*, 13 (2), 247–266.
Cook, T. and Hess, E., 2007. What the camera sees and from whose perspective: fun methodologies for engaging children in enlightening adults. *Childhood*, 14 (1), 29–45.
Cooke, B. and Kothari, U., 2001. *Participation: the new tyranny?* London: Zed Books.
Crozier, G. and Davies, J., 2007. Hard to reach parents or hard to reach schools? A discussion of home-school relations, with particular reference to Bangladeshi and Pakistani parents. *British educational research journal*, 33 (3), 295–313.
David, T., ed., 1994. *Working together for young children.* London: Routledge.
Department of Health, 2003. *Getting the right start. The national service framework for children. Emerging findings.* London: Department of Health, HMSO.
DFE, 1994. *Code of practice on the identification and assessment of special educational needs.* London: DFE.
DFEE, 2000. *SEN code of practice on the identification and assessment of pupils with special educational needs & SEN thresholds: good practice guidance on identification and provision for pupils with special educational needs.* London: DFEE.
DfES, 2001. *SEN toolkit. Section 4. Enabling pupil participation.* London: DfES.
DfES, 2003. *Every child matters.* London: HMSO.
DfES, 2004. *Every child matters and every young person. What you said and what we're going to do.* London: DfES.
Flicker, S. and Guta, A., 2008. Ethical approaches to adolescent participation in sexual health research. *Journal of adolescent health*, 42 (1), 3–10.
Fundudis, T., 2003. Consent issues in medico-legal procedures: how competent are children to make their own decisions? *Child and adolescent mental health*, 8 (1), 18–22.
Gallagher, K., ed., 2008. *The methodological dilemma. Creative, critical and collaborative approaches to research.* Abingdon: Sage.
Galloway, D., Armstrong, D., and Tomlinson, S., 1994. *The assessment of special educational needs Whose Problem?* Harlow: Longman.
Grossman, V., 2005. *An investigation into the roles children have in educational psychology practice.* MSc in Educational Psychology Dissertation. Newcastle University.
Guillemin, M. and Gillam, L., 2004. Ethics, reflexivity, and "ethically important moments" in research. *Qualitative inquiry*, 10 (2), 261–280.
Hacking, E.B., Barratt, R., and Scott, W., 2007a. Engaging children: research issues around participation and environmental learning. *Environmental educational research*, 13 (4), 529–544.
Hacking, E.B., Scott, W., and Barratt, R., 2007b. Children's research into their local environment: Stevenson's gap, and possibilities for the curriculum. *Environmental education research*, 13 (2), 225–244.
Hart, R.A., 1992. *Children's participation: from tokenism to citizenship.* Florence: UNICEF.
HMSO, 1989. *The children act 1989.* London: HMSO.
Hobbs, C., 2000. *Initial consultation with children and young people: the practice of educational psychologists. DEdPsy assignment.* Newcastle: Newcastle university.
Hobbs, C., et al., in press. The professional practice of educational psychologists: developing narrative approaches. *Educational and child psychology*, 29 (1).
Hughes, R., 2002. *The voice of the child within the family group conferencing process.* MSc in Educational Psychology Dissertation. Newcastle University.
Investing in Children, 2011. *Investing in Children: Annual Report.* Durham: Investing in Children.
Jackson, A.Y. and Mazzei, L.A., eds., 2009. *Voice in qualitative inquiry.* Abingdon: Routledge.
James, A., 2007. Giving voice to children's voices: practices and problems, pitfalls and potentials. *American anthropologist*, 109 (2), 261–272.
James, A. and Prout, A., 1990. *Constructing and reconstructing childhood. contemporary issues in the sociological study of childhood.* London: Falmer Press.
Kaplan, I., Lewis, I., and Mumba, P., 2007. Picturing global educational inclusion? Looking and thinking across students' photographs from the UK, Zambia and Indonesia. *Journal of research in special educational needs*, 7 (1), 23–35.

Kellett, M., 2010. *Rethinking children and research*. London: Continuum.

Kesby, M., 2007. Spatialising participatory approaches: the contribution of geography to a mature debate. *Environment and planning*, 39 (12), 2813–2831.

Leitch, R., et al., 2007. Consulting pupils in assessment for learning classrooms: the twists and turns of working with students as co-researchers. *Educational action research*, 15 (3), 459–478.

Lewin, K., 1951. *Field theory in social science: selected theoretical papers* (D. Cartwright, ed.). New York: Harper and Row.

Lightfoot, J. and Sloper, P., 2003. Having a say in health: involving disabled and chronically ill children and young people in health service development. *Children and society*, 17 (4), 277–290.

Marquez-Zenkov, K., et al., 2007. If they'll listen to us about life, we'll listen to them about school: seeing city students' ideas about 'quality' teachers. *Educational action research*, 15 (3), 403–415.

McClelland, S.I. and Fine, M., 2008. Writing on celephane. *In*: K. Gallagher, ed. *The methodological dilemma. Creative, critical and collaborative approaches to research*. Abingdon: Routledge.

McDowell, L., et al., 2005. The contradictions and intersections of class and gender in a global city: placing working women's lives on the research agenda. *Environment and planning*, 37 (3), 441–461.

Mentoring and Befriending Foundation, 2010. *Peer mentoring in schools*. Manchester: Mentoring and Befriending Foundation.

Moore, J. and Dunworth, F., 2011. Review of Evidence from Aimhigher Area Partnerships of the Impact of Aimhigher January 2011. Aim Higher. Available from: http://www.aimhigher.ac.uk/sites/practitioner/resources/Evidence%20Report%2031%20January%202011.pdf [Accessed 11 March 2012].

Moore, R. and Muller, J., 1999. The discourse of 'voice' and the problem of knowledge and identity in the sociology of education. *British journal of sociology of education*, 20 (2), 189–206.

Morrow, V., 2008. Ethical dilemmas in research with children and young people about their social environments. *Children's geographies*, 6 (1), 49–61.

Moss, P., 2010. Foreward. *In*: A. Clark, ed. *Transforming children's spaces. Children's and adults' participation in designing learning environments*. London: Routledge.

Murdoch, J. and Pratt, A.C., 1993. Rural studies: modernism, postmodernism and the 'post-rural'. *Journal of rural studies*, 9 (4), 411–427.

Nystrom, E., 2007. Exclusion in an inclusive action research project: drawing on student perspectives of school science to identify discourses of exclusion. *Educational action research*, 15 (3), 417–440.

Ofsted. 2003. *Inspecting schools. Framework for inspecting schools*. Effective from September 2003. London: Ofsted.

Philo, C., 1993. Reply to comment. Postmodern rural geography? A reply to Murdoch and Pratt. *Journal of rural studies*, 9 (4), 429–436.

Punch, S., 2002. Research with children. The same or different from research with adults? *Childhood*, 9 (3), 321–341.

Reay, D., 2004. Educational and cultural capital: the implications of changing trends in education policies. *Cultural trends*, 13 (50), 73–86.

Reay, D., 2006. 'I'm not seen as one of the clever children': consulting primary school pupils about the social conditions of learning. *Educational review*, 58 (2), 171–181.

Robson, E., et al., 2009. 'Doing it right?': working with young researchers in Malawi to investigate children, transport and mobility. *Children's geographies*, 7 (4), 467–480.

Sandbaek, M., 1999. Children with problems: focusing on everyday life. *Children and society*, 13 (2), 106–118.

Schäfer, N. and Yarwood, R., 2008. Involving young people as researchers: uncovering multiple power relations among youths. *Children's geographies*, 6 (2), 121–135.

Shemmings, D., 2000. Professionals' attitudes to children's participation in decision-making: dichotomous accounts and doctrinal contests. *Child and family social work*, 5 (3), 235–243.

Shier, H., 2001. Pathways to participation: openings, opportunities and obligations: a new model for enhancing children's participation in decision-making in line with article 12.1 of the United Nations Convention on the Rights of the Child. *Children and Society*, 15, 107–117.

Sin, C. and Fong, J., 2010. Commissioning research, promoting equality: reflections on the disability rights Commission's experiences of involving disabled children and young people. *Children's geographies*, 8 (1), 9–24.

Thomson, P., ed., 2008. *Doing visual research with children and young people*. Abdingdon: Routledge.

Thomson, P. and Gunter, H., 2006. 'From "consulting pupils" to "pupils as researchers": a situated case narrative'. *British educational research journal*, 32 (6), 839–856.

Thomson, P. and Gunter, H., 2007. The methodology of students-as-researchers: Valuing and using experience and expertise to develop methods. *Discourse: studies in the cultural politics of education*, 28 (3), 327–342.

Todd, L., 2000. Letting the voice of the child challenge the narrative of professional practice. *Dulwich centre journal*, 1 & 2, 73–79.

Todd, L., 2007. *Partnerships for inclusive education: a critical approach to collaborative working*. London: Routledge.

Tolley, E., et al., 1998. *Young opinions great ideas*. London: National Children's Bureau.

Vanderbeck, R., 2008. Reaching critical mass? Theory, politics, and the culture of debate in children's geographies. *Area*, 40 (3), 393–400.

White, M., 2007. *Maps of narrative practice*. London: Norton.

Whitty, G. and Wisby, E., 2008. Whose voice? An exploration of the current policy interest in pupil involvement in school decision-making. *International studies in sociology of education*, 17 (3), 303–319.

Williamson, B., 2003. *Grit in the Oyster: Final report of the evaluation of investing in children*. Durham University, Durham. Available from: http://www.iic-uk.org/modules/download_gallery/dlc.php?file=59 [Accessed 11 March 2012].

Williamson, B. and Cairns, L., 2005. *Working in partnership with young people: from practice to theory*. Durham: Investing in Children & Research in Practice.

Williamson, B. and Cairns, L., 2006. *Working in partnership with children, young people and their families*, Report of a Conference in Cheadle, 21 February 2006. Investing in Children & Research in Practice: Durham.

What we say and what we do: reflexivity, emotions and power in children and young people's participation

Victoria Jupp Kina

School of Health Sciences and Social Care, Brunel University, Kingston Lane, Uxbridge, UB8 3PH, UK

> Children and young people's participation is now recognised as a relational process and consequently the role of adults has been identified as crucial. Yet the role that adults play in the participatory process is underexplored and how underlying factors of emotions and power influence this role is unclear. Using the work of Spinoza as a framework for discussion, this article explores the impact of emotions and power on participatory processes. Based on doctoral participatory action research undertaken in São Paulo, Brazil, I argue that adults facilitating participation need to reflect on how they are personally implicated in the participatory process if underlying factors of emotions and power are to be acknowledged and addressed.

The role of emotions and power in participation

It is now broadly accepted that participation is a relational process that cannot be understood outside the set of relationships that constitute all the actors (Prout and Tisdall 2006). Consequently, the role of adults has been identified as crucial to the development of children and young people's participation (Percy-Smith and Thomas 2010). Heavily influenced by the recent return to dialogical pedagogy (Freire 1996, 2001), situated and partial knowledges (Rose 1997, Laurie et al. 1999) and action research (Pain 2004, Cahill 2007, Percy-Smith 2007), there is now increasing recognition of participation as 'a partial, situated and contestable work-in-progress, subject to future challenge and transformation of *all* parties involved, with effects felt both locally and in more distant contexts' (Mannion 2010, p. 338). This has further reinforced the challenging of the traditional oppositional positioning of children and adults through the new social studies of childhood and the ensuing recognition of children as social agents (Prout and James 1997). This has also led to wider acceptance of the view that both adults and children need to be considered as 'becomings' (Uprichard 2008).

Yet despite calls to 'shift attention from children *per se* to children in relation to others' (Prout and Tisdall 2006, p. 243), the role that adults play in supporting children and young people's participation, both in research and practice, remains an area that is relatively overlooked (Wyness 2009). Some valuable recent contributions highlighting the complexity of children and young people's participation have emerged from the human geography field (Kraftl and Horton 2007, Percy-Smith 2007). These have helped in breaking down the adult/child dichotomy and shifting debates to recognising the importance of the everyday nature of participatory practices. Yet, while these contributions acknowledge the complexity of the adults' role, an explicit exploration of this has still to be fully developed. Further to this, while issues of emotions and power have been recognised as central aspects of participation, exactly how they impact on the participatory process remains unclear. While the situated nature of our knowledges has been extensively explored (England 1994, Rose 1997, Laurie *et al.* 1999), this has focused on the need to recognise our political positionality – through our gender, ethnicity, class, age or (dis)ability – and recent authors have argued that such understanding of situated knowledge is limited, highlighting the need to recognise the relational dimension of knowledge and the impact of the highly subjective nature of knowing, for example, through the impact of our ability to relate to those around us through our personality (Moser 2008). This has been a valuable contribution to the debates on the subjective and subtle ways that both emotion and power are played out within participatory processes within both research and practice.

This article aims to build on these contributions and further our understanding of the relational nature of children and young people's participation through exploring the impact of emotions and power on the roles played by adults in participatory processes with children and young people. Based on doctoral research undertaken in São Paulo, Brazil, over a period of 1 year, this article explores how emotions and power relations underlying participatory processes can be made visible. Drawing on the use of reflexivity in feminist writing (Rose 1997, Moser 2008, Askins 2009, Nicholls 2009), I argue that adults facilitating participation need to reflect on how they are personally implicated in the participatory process if underlying factors of emotion and power are to be acknowledged and addressed.

The role of emotions: a Spinozian view

A key area of children and young people's participation is the impact of emotions on participatory processes. While emotions have been explored in the literature on organisational change through the development of the notion of 'emotional intelligence' (see, e.g. Goleman 2001), the role of emotions and emotionality has received relatively little attention in relation to children and young people's participation (Pinkey 2011). Emotions have begun to be more fully explored in the literature on participatory research, particularly within the field of feminist geography (Widdowfield 2000, Jupp 2008, Moser 2008, Askins 2009, Punch 2011), and I will draw on this thinking throughout this article. However, in exploring the role of emotions and power in participatory processes, I also draw on the ideas of the seventeenth century philosopher Spinoza as a framework for discussion. There has recently been a return to Spinoza's philosophy with his work being taken up by both Deluze and Lukes to inform their thinking on power (Moreau 1996, Lukes 2005) and within debates on emotions, particularly within emotional and affective geographies (Pile 2010, 2011, Curti *et al.* 2011, Dawney 2011). Spinoza's concept of potential ('*potentia*' in Latin) was also extensively explored by Vygotsky and introduced into psychology discourses to overcome the discipline's negative view of the relevance of emotions, which were at the time considered antagonistic to reason (Sawaia 2009). The use of Spinoza's ideas as a framework to explore both power and the emotions offers a fruitful avenue to improve our understanding of the impact of emotions and power on children and young people's participation. Moving away from Cartesian conceptualisations that separate thought and emotion towards a more inclusive conceptualisation that accepts the emotions as both fundamental and necessary, this article aims to bring some clarity to what remains a rather cloudy area of debate.

The revolutionary aspect of Spinoza's philosophy was his conceptualisation that the body and the mind are of one substance; that it is through the human body being affected that the human mind is able to recognise the body's existence. Crucially, he challenged the previous Cartesian conceptualisations that valued the primacy of the mind, arguing that the power of thought and action 'emerge from a process of sustained engagement with the world' (Brown and Stenner 2001, p. 86). In other words, it is through our engagement with the world, and how we are affected as a result of this engagement, that allows for both thought and action. Vygotsky's application of Spinoza's philosophy led him to conclude that 'the separation of the intellectual side of our consciousness from the affective-volitional side constitutes one of the most serious basic defects of all traditional psychology' (Vygotsky 1993 cited in Sawaia 2009 [my translation]). For both Spinoza and Vygotsky, thoughts are based on our consciousness which is, in turn, a result of our needs, interests, impulses and emotions. In short, our emotions are inseparable from thought and what we think is essentially inseparable from what we feel.

Interestingly, however, the shift away from Cartesian understandings that separate mind and body – and therefore thought and emotion – continues to prove challenging. As clearly highlighted by the recent response to Pile's (2010) discussion of affective and emotional geographies (see Curti *et al.* 2011, Dawney 2011, Pile 2011), the articulation of Spinoza's philosophy is challenging and the interpretation and application of his ideas varied. Within this article, I do not wish to engage with the debates of affective and emotional geographies, partly due to a deep discomfort with the dichotomous effect that such division has created. This division between affect and emotion and the ensuing maintenance of the distinction between emotion and reason has been discussed elsewhere (see Thien 2005). In my view, this division shifts us away from the Spinozian view that mind and body are of one substance and undermines the fundamental role that emotions play within everyday experience. I do not view emotions and affect as separate entities, but instead agree with the view extended by Ahmed (2004) and Askins (2009) that affect and emotion are intrinsically interrelated:

> ...emotions are how I respond (how I care about/show interest in something); affect is what intensifies my response (or my capacity to be affected). Both are physiological (preconsciously embodied) *and* socially circulating; both are caught up in past associations/experiences/histories, *and* dynamic and emergent. (Askins 2009, p. 10 [emphasis in original])

It is this understanding of emotion that underpins this article, yet my choice to use the term 'emotion' rather than 'affect' is also a political one. The aim of this article is to explore people's lived experiences in order to improve our understanding of our own actions, and to do this, it is important to engage with people's own experiences. Our lived experiences are framed by emotions and therefore, while my theoretical understanding of emotions is underpinned by the notion of affect, I wish to engage with people's lived experiences as they experience them. As noted by Lupton (1998, p. 6):

> ...our concepts of our emotions are often integral to our wider conception of our selves, used to give meaning and provide explanation for our lives, for why we respond to life events, other people, material artefacts and places in certain ways, why we might tend to follow patterns of behaviour throughout our lives.

Alongside his arguments on affect, an important element of Spinoza's thinking that can help to develop a more nuanced understanding of participation is his development of the idea of '*potentia*' or potential. For Spinoza, all people are born free and to be free is to be 'potent', or have the potential to act. But this potential increases or decreases depending on experience of encounters with other bodies. Poor experiences of encounters that limit potential to act are, according to Spinoza, based on 'inadequate ideas'. By contrast, positive encounters are based on 'adequate ideas' that satisfy all aspects of being human. For an idea to be adequate, it needs to be something that is truly required, satisfying all aspects of being human. Inadequate ideas lead to passive actions, or 'passions', while adequate ideas lead to active action as they are based on both an explicit awareness of not only need but the reasons for that need.

Background to the study: research design and methods

I made use of the connections between adequate and inadequate ideas and passive and active action in a participatory action research project with three NGOs in São Paulo, Brazil. I worked alongside NGO staff to plan, implement and reflect upon a range of participatory methods with children and young people attending activities offered by the NGOs with the aim of unravelling some of the complexities of participatory practice. The objective of the research was to recognise the relational nature of participatory process and use praxis – the process of reflection and action – to explicitly analyse the role of the adult within children and young people's participation. The motivations for undertaking this research lie in previous frustrations within the international development sector in general and with my observation of numerous failures to develop children and young people's participation in particular. It was while qualifying as a social worker that I came into contact with theoretical writing about power and it was through this I began to reflect on these experiences and to place them within a theoretical framework. When the opportunity to study for a PhD emerged, I decided to use the opportunity to explore these frustrations in more detail to try to improve understanding of what I could see was a complex and often misunderstood process.

Although I worked with three NGOs during my fieldwork, this article draws on the research experience at just one of the NGOs. This NGO was located in a municipality on the periphery of São Paulo city with the highest population density in the state of São Paulo. The area is characterised by precarious housing conditions, higher levels of violence than found in other municipalities and low-income levels. The NGO's multidisciplinary team aims to minimise factors leading to social exclusion through group and individual support to young people and their families, alongside providing educational and recreational workshops and community services. At the time of the research, the organisation had five main areas of work that were structured in five 'nuclei': the nucleus of social work, the nucleus of education, the nucleus of community development, the youth protagonism nucleus and a research nucleus. The work is wide-ranging, including child protection interventions, educational activities for children and young people of all ages, support for community development projects and business initiatives, as well as the provision of community services such as a community library. This NGO also has a system of employing young people from the local community as youth mediators. These young staff members are between 15 and 18 years of age and are responsible for the day-to-day running of the community library, the coordination of activities within local schools and also the development of community development initiatives and campaigns in the local area.

The research process at this NGO included 27 semi-structured interviews with all staff at the beginning and end of the project (53 interviews in total) and participant observation. In the latter, I actively engaged with my own experiences while working alongside the staff and young people through the use of both written and videoed fieldwork diaries. I also facilitated three participatory workshops with all staff members, including young people employed as youth mediators (there was an average of 33 participants in each workshop), to encourage group reflection on the development of children and young people's participation within the organisation. Interviews, conducted in Portuguese, lasted between 20 min and 2 h. They were audio recorded and then transcribed, analysed and coded in Portuguese, using Nvivo. A modified version of grounded theory was used throughout the fieldwork that, while not aiming for saturation, made use of personal reflections and memos that ensured the 'initial freshness of the analyst's theoretical notions' (Glaser 1965, p. 439).

Accepting the emotions: the need for coherence

The following extract is taken from a final interview with Paula, a nucleus coordinator:

> Eu acho que o maior desafio é as pessoas encararem isso, a participação, como estratégias que fazem com que elas aprendam habilidades pessoais. Como essa estratégia é uma estratégia que mobiliza muitos recursos internos, as pessoas não vão querer, elas vão ter mais resistência a criar estratégias ou a melhorar, ou sair do modelo que elas conhecem, porque elas não aguentam o que elas encaram.
>
> I think the biggest challenge is people facing up to this, participation, as strategies that make you learn personal skills. As this strategy is a strategy that mobilises a lot of internal resources, people won't want it, they will have more resistance to create strategies or to improve, or to leave the model that they know, because they can't bear what they face. (Paula, staff member)

While reflecting on the research process during her final interview, Paula clearly demonstrates her view of participation as a deeply personal process. It involves strategies that utilise 'internal resources' and focuses on developing personal skills. This involves people being both willing and able to take an honest look at themselves, to identify and acknowledge their weaknesses and seek to address them. Paula indicates her understanding of participation as more than an intellectual decision, but also as a commitment to personal change. It requires a commitment that goes far beyond believing that participation is a good thing, a way to create positive change or challenge current social structures or inequality. Scott-Villiers (2004) has argued that effective participation requires each of us to make an effort to understand what is inside ourselves and to achieve this 'we must start with investigating our own philosophies' (p. 205). While the belief in participation as a 'good thing' is clearly a starting point for many participation practitioners, this remains an intellectual decision, reinforcing the binary distinctions of us/them and self/other as the process focuses on 'them'. This more limited understanding of participation has been strongly criticised within the international development field (see Cooke and Kothari 2001) and the argument that such dualisms actually work to reinforce privilege has been comprehensively made, as whilst promoting the 'other's' empowerment, our own actions hinge on '*our* complicity and desire' (Kapoor 2005, p. 1204). As Paula's words indicate, the biggest challenge for participatory practice is not about how to include 'them', it is about being prepared to include *ourselves* within the process; it is about us being prepared to face up to our own limits.

Yet despite the increasing acceptance of the need to move beyond the dualisms of us/them and self/other to include ourselves within the participatory process, this is far from simple. As has been effectively highlighted within feminist geography through the unravelling of some of the complexities of reflexivity, the inclusion of self is a deeply challenging process. As noted by Punch (2011) in her discussion of emotions in research, confronting our failings and opening up our feelings to public scrutiny is no easy task. Yet the failure to include the subtler aspects of participation while still trying to 'achieve' it overlooks the relational nature of the process. As argued by Jupp (2008), it is important that we remember the value of everyday, relational interactions and feelings as constitutive of spaces of participation, and Kraftl and Horton (2007) suggest that participation is a sense or a feeling that can emerge out of relatively ordinary situations. Mannion (2010, p. 338) has recently argued that 'current versions of children and young people's participation offer a limited view of the ways in which participants and spaces reciprocally trigger changes in each other' and I argue that this limited view results from the decision to 'be participatory' remaining an intellectual decision that fails to recognise the relational through making the commitment to personal change. This results in practitioners being trapped in dangerous territory where pre-existing relationships can all too easily be reproduced rather than challenged. To quote from Freire (1996, p. 48), 'Political action on the side of the oppressed must be pedagogical action in the authentic sense of the word, and, therefore, action *with* the oppressed'. To take action *with* the oppressed requires placing yourself as a subject of the process alongside those with whom you are working. While deeply challenging – and potentially deeply uncomfortable – it is only through accepting this aspect of the process that it is possible to avoid making people 'the objects of ... humanitarianism' which 'itself maintains and embodies oppression' (Freire 1996, p. 36). Anything less traps practitioners into acts of false generosity that reinforce rather than challenge privilege. After all, as Kapoor (2005, p. 1207) argues: 'Pretending to step down from power and privilege ... is a reinforcement, not a diminishment, of such power and privilege'.

Yet while knowing what we should do and actually feeling able to act on this may indeed be a complex process, if we return to the ideas of Spinoza that emotions are inseparable from thought, it becomes clear that without understanding why we do what we do or think how we think, in other words, how our emotions impact upon our thinking and actions, we run the risk of incoherence in our actions. To illustrate my thinking, I have developed the metaphor of the ocean. Figure 1 shows that the ocean surface represents our actions, the visible aspect of the participatory process – what we do. Underneath the ocean surface there are two currents. The top current represents our intellectual level of understanding – the way we think and the decisions we make. This level has a visible impact on what we see at the ocean's surface. However, a deeper current at the ocean bed represents our emotional level of understanding – what we feel. This current is less visible from the ocean surface. It can be seen but its influence is often blocked by the stronger current above it. I argue that each current is in dialogical relationship with the other and that while we need to view the ocean as one whole – after all, as Spinoza highlights thought and emotion are essentially inseparable – we also need to recognise how each part of the ocean relates and impacts on the other to understand the conditions within the ocean as a whole. For example, the longer the current at the ocean bed – what we feel – is blocked by the current above it – what we think – the more unsettled, opaque and inconsistent the ocean surface will appear. And it is this, the inconsistency of the ocean surface as a result of the lack of coherence between the two currents underneath it, which can help us to understand the difficulties of the participatory process.

But this raises the question of how we can begin the process of understanding our own 'oceans'. At this point, it is useful to turn to the literature on critical reflexivity. Critical reflection and researcher reflexivity are advocated particularly within feminist geography as a means to situate our knowledge and avoid the 'false neutrality and universality of so much academic knowledge' (Rose 1997). Critical reflection has also been advocated in research as a means by which to address 'the presence of the knower in the known and vice versa' (Gray 2008, p. 936). Yet critical reflection in research has been criticised as lacking a clear conceptual base

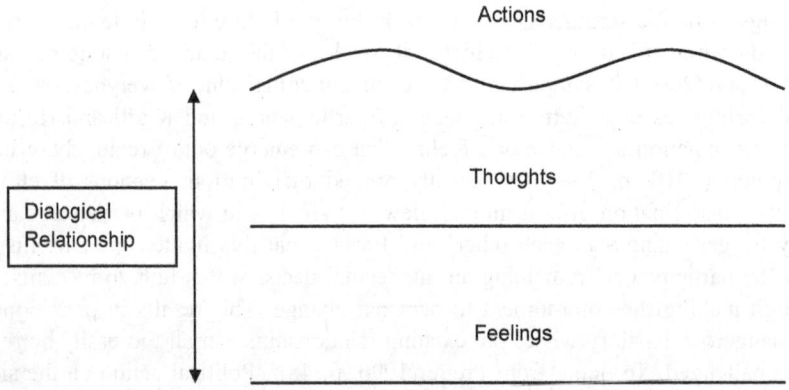

Figure 1. The dialogical relationship between what we feel, think and do.

and a clear understanding of what this means in practice (Chiu 2006). Chiu (2006) and Reason and Torbert (2001) proposed that there are at least three layers of reflection, conceptualised as first, second and third person inquiry, that offer a more holistic view of research that recognises three interrelated audiences of the research process:

> All good research is for me, for us, and for them: it speaks to three audiences . . . It is for them to the extent that it produces some kind of generalizable ideas and outcomes . . . It is for us to the extent that it responds to concerns for our praxis, is relevant and timely . . . [for] those who are struggling with problems in their field of action. It is for me to the extent that the process and outcomes respond directly to the individual researcher's being-in-the-world. (Marshall and Reason 1994 cited in Reason and Torbert 2001)

These have also been further developed by Nicholls (2009) as self-reflexivity, inter-personal reflexivity and collective reflexivity. Both Marshall and Reason (1994 cited in Reason and Torbert 2001) and Nicholls (2009) effectively highlight the complex nature of critical reflection. However, both of these approaches are based on the concept of knowledge as hierarchical which goes against Spinoza's concept of the body and mind being of one substance. I argue that there is an underlying factor that connects all three layers of the reflective process, the recognition of our emotions, and it is this that determines our ability to incorporate these layers into our practice. As Gray (2008, p. 936) has argued, emotion in the practice of reflexivity tends to be overlooked despite it being a crucial aspect of the method precisely because 'affectivity is a way of apprehending the world'. So while it is essential that we understand the different forms of reflexivity, it is equally important that we understand what underlies them to work out how they can be developed and strengthened.

Paula's reflections – they cannot bear what they have to confront – offer us some empirical insight into why emotion in both participatory and reflective processes have proved such challenging processes to undertake. Put simply, reflective processes imply significant emotional demands. While participatory processes require each person to undertake first, second and third person reflections, an underlying question is whether the individual is both able and willing to address their own emotions. As noted recently by Moser (2008, p. 389):

> People's different emotional intelligence or personalities make some people more able than others to recognize when their personality is hindering or affecting their research.

It is of fundamental importance that we do not underestimate the impact of these subtler aspects on the participatory process. As clearly articulated by one of the staff members in this study, the relational nature of the process and the ability of the facilitator of a participatory process to relate to the participants fundamentally impacts on their willingness to participate fully in the process:

Victoria: Vocês acham que esse trabalho deve continuar depois? Vocês querem fazer mais reuniões assim?

Suelita: Fazer mais, mas que tenha uma pessoa que explique direito e tenha paciência. Tem tudo isso, não adianta querer fazer uma coisa que você não tem paciência de ensinar, que não gosta que a gente pergunte. Aí não adianta, nem precisa continuar.

Victoria: Tem gente que não gosta de perguntar?

Suelita: Aí não precisa continuar, deixa como está. Se vir uma pessoa legal, que ensine, que explique, que a gente possa entender, que a gente possa perguntar, pode ter bastante. Quanto mais melhor. Mas uma pessoa que não vai ensinar direito, que não vai ter paciência e que não fica claro, nem adianta continuar porque não estou nessa, entende?

Victoria: Para vocês uma das coisas mais importante é a pessoa, o jeito da pessoa que está fazendo, para ter essas qualidades que vocês falaram?

Suelita: Que tenha paciência, o respeito, entendeu? Não ter você que está explicando e a Tanda que é uma voluntaria perguntar para você, e você não dar atenção porque a Tanda é voluntaria 'não vai entender nada'. Não vai adiantar, aí é melhor nem me por nos grupos porque vou participar e não vou participar de corpo e alma, vai participar só o espírito e uma obrigação. Eu não vou me interessar. Eu sou muito assim, se vejo que aquela pessoa não me dá bola, eu também não dou bola para ela. Pode ser o rei da Inglaterra, se eu vejo que não me respeita, eu também não respeito. Eu não vou desrespeitar com palavras, falar 'oi, tudo bem', tenho que fazer as minhas obrigações, faço e pronto, mas só por obrigação.

Victoria: Do you think that this work should continue? Do you want to have more meetings like this?

Suelita: Do more, but when there is someone who explains clearly and has patience. You have all this, but it doesn't help wanting to do something if you don't have the patience to teach, if you don't like that we ask questions. It doesn't help, there's no need to continue.

Victoria: There are people that don't like to ask questions?

Suelita: There's no need to continue, leave it as it is. If someone cool comes, that teaches, that explains, that we can understand, who we can ask questions, you can have loads. More the better. But someone that won't teach properly, that won't have the patience and that isn't clear, it doesn't help to continue because I'm not in this, you see?

Victoria: So for you, one of the most important things is the person, the skill of the person that's facilitating, to have these qualities that you spoke of?

Suelita: That has patience, respect, understand? Not you explaining and Tanda who is a volunteer asking you something, and you don't pay attention because Tanda is a volunteer 'she won't understand anything'. It won't help, it's better to not put me in the groups because I'll participate and I won't participate with body and soul, I'll participate only in spirit as an obligation. I won't be interested. I'm very much like that, if I see that someone doesn't give me the ball, I also won't give the ball to them. It could be the Queen of England, if I see that she doesn't respect me, I also won't respect her. I won't disrespect with words, I say 'hi, how are you', I have to meet my obligations, I'll meet them and done, but only as an obligation. (Suelita, staff member)

Suelita clearly articulates the impact that the ability of an individual to relate to the people with whom they are working can have on a participatory process. I argue that this ability to relate to others is fundamentally affected by our ability to relate to ourselves. Returning to Spinoza, if addressing our emotions requires the recognition of how we both affect and are affected by our engagement with the world, then it seems the crucial aspect that underlies both participation and reflexivity is the ability and willingness of each of us to recognise this engagement. Therefore, whilst participation has been recognised as a potential process of transformation (Kesby 2005), I argue that this can only occur if all those involved in the process both *wish and are able to be transformed*.

Making the invisible visible: the role of the participatory process in addressing emotion and power

Throughout the research process, it became clear that ability and willingness to recognise our affective engagement is not always possible. Feeling tired, nervous, excited, stressed, disappointed or angry, for example, can all too easily threaten our capacity for understanding the currents of our personal oceans. However, what also became clear was that the use of a participatory process to develop children and young people's participation created an opportunity to make the invisible visible. It allowed the participants to move below the top current towards the ocean floor, to get beyond the intellectual level of what we think and recognise our own subjectivity, to begin to discuss not just what we think, but why we think so, and how this relates to what we feel.

For example, the act of discussing children and young people's participation, and the use of a participatory approach within these discussions, created an opportunity for self-reflection and participants began to connect their own inconsistencies and the inconsistencies of other people to the uneven development of the participatory process. This was particularly true for some of the young staff members employed as youth mediators. For example, during final reflections on the research process one of the youth mediators, Thiago, describes a process of personal 'unblocking' whereby he overcame his lack of confidence to express his opinion:

> Eu me sentia bloqueado antes. Eu acho que não conseguia expressar tudo que eu pensava e falava 'ai, você vai falar besteira, por que eu vou falar? Deixa eu ficar calado porque ganho mais e aprendo mais ouvindo mesmo'. Mas não, entendi que é falando que a gente aprende mesmo, que é participando que a gente aprende, né? E é isso que eu estou tentando fazer até hoje.

> I felt blocked before. I think that I wasn't able to express all that I thought and I said [to myself] 'ah, you'll talk crap, why say anything? Keep quiet because I'll get more [out of it] and learn more just from listening.' But no, I understood that it's through speaking that we learn, that it's through participating that we learn, isn't it? And it's this that I'm trying to do today. (Thiago, youth mediator)

What was so interesting about Thiago's reflections was that he states that he felt he had been blocked 'before', meaning before he participated in the research. Thiago relates this unblocking to his participation in the staff workshops. When asked what he had felt was positive about the research, Thiago responded:

> Thiago: Acho que foi bom porque comecei a ter umas ideias que pude usar não só na biblioteca, mas para mim, para os meus próprios conhecimentos. Eu acho que me ajudou bastante.

> Victoria: Como ajudou você? Fora do seu trabalho?

> Thiago: Isso. Nos meus estudos. O meu quarto é a minha biblioteca, fico lá estudando, pesquisando alguma coisa e a participação me ajuda de uma forma não ... Não vai ter ninguém para participar junto, mas participando comigo mesmo. Não sei se você consegue entender, mas acho que me ajuda um pouco esse negócio de poder expressar o que eu sinto. Antes eu não me expressava, depois da primeira oficina que a gente teve com você comecei a ter um pouquinho mais de ideias, tenho participado. Eu acho parte de uma equipe. Eu acho que me ajudou bastante.

> Thiago: I think it was good because I started to have ideas that I could use not just in the library but for me, for my own knowledge. I think it helped me a lot.

> Victoria: How did it help you? Outside of your work?

> Thiago: That's it. In my studies. My bedroom is my study, I stay there studying, researching something and participation helps me in a way, not ... There won't be anyone to participate with, but participating with myself. I don't know if you're able to understand, but I think it helped me a little with this thing of being able to express what I feel. Before I didn't express myself, after the first workshop that we had with you I started to have a few more ideas, having participated. I felt part of the team. I think it helped me a lot. (Thiago, youth mediator)

If we briefly return to the metaphor of the ocean, in essence what Thiago is highlighting is that, through his participation in the staff workshops, his intellectual and emotional currents appeared to become more coherent. He had noticed a subtle change within himself in that he had begun to understand why he felt unable or unwilling to participate previously and as a result had started participating 'with himself', which also led to him feeling more like he was part of the team. In Spinoza's terms, it appears that Thiago feels he moved from passive to active action as he had begun to understand not only what he felt, but also why he felt it. This move from passive to active action was also clearly highlighted during final reflections with Susana, another youth mediator at the NGO. When asked if her understanding of participation had changed during the research Susana replied:

> Sim. Porque o meu pensamento a respeito de participação era completamente diferente. Para mim, participação só era quando uma pessoa falava ou como ... mas com essas oficinas percebi que participação vai muito além do que falar, mas traduzida em palavras escritas, gestos e até mesmo símbolos.

> Yes. Because my thinking with regards to participation was completely different. For me, participation was only when a person spoke or how ... but with these workshops I realised that participation goes much further than speaking, but translates into written words, gestures and even symbols. (Susana, youth mediator)

As the interview progressed, it became clear how this change in her understanding of participation had also influenced Susana to change her behaviour and develop new ways of communicating with people. When asked if the research had had an impact on her work, Susana responded:

> Eu vou participar do meu jeito agora. Eu vou escrever, ficar mandando para as pessoas o que estou escrevendo, até me desenvolver totalmente para falar em público.

> I'll participate my way now, I'll write, keep sending people what I'm writing, until I completely develop [have enough confidence] to speak in public. (Susana, youth mediator)

This highlights the fundamental connection between what we think and what we feel. Susana's improved understanding of what participation involves helped her to participate in 'her way'. I worked closely with Susana during the research process and although at 19 years of age she was one of the older 'young people' working at the NGO, she was one of the shyer members of the younger staff. Although Susana recognised her own personal development since she had started working at the NGO and her increased confidence, she was still extremely nervous when talking in front of groups of people or with people with whom she did not feel comfortable. On two occasions during the research process, Susana wrote down her opinions and reflections in the form of a letter in order to express her frustration about certain events that had occurred. She recognised that she is both more comfortable and expresses herself better through the written word rather than verbal dialogue, but she had never previously expressed her views in this form. This form of expression had come completely from Susana – I had made no requests for any form of written feedback nor explicitly stated writing as an alternative option. The fact that Susana had found an alternative way to express herself indicates that she, like Thiago, had also become more active in her own participation.

The impact of the participatory research process was also highlighted during the final reflections of Fabio, a nucleus coordinator at the NGO. He identified how the research process affected people both professionally and personally in different ways:

> Muitos extremamente interessados, interesse não só profissional mas também pessoal assim, em relação com a vida, com o que acha que tem que ser, como a própria vida tem que ser. E outras incomodadas, incomodadas no mesmo sentido que tem a ver consigo, com a instituição, com a vida e tudo mais. São as coisas que não dá para ver, não dá para sentir, mas que mexeu bastante ... na pesquisa teve essa influência que eu vejo como positiva. Sei lá, não acho que vai resolver o problema mas deu

uma, transpareceu, apareceu algumas coisas de algumas pessoas, de alguns grupos que fazem as outras pessoas entenderem melhor o que é real mesmo.

Many were extremely interested, not just a professional interest but a personal one as well, in relation to life, with what I think has to be, how my own life has to be. And others were uncomfortable, uncomfortable in the same sense that it has to do with you, with the institution, with life and all that. They're the things that you can't see, can't feel, but that mess with you a lot ... the research had this influence that I see as a positive. I don't know, I don't think it'll resolve the problem but it gave a, transparency, a few things appeared in a few people, in a few groups that made other people understand better what is actually real. (Fabio, staff member)

During her final interview, Beatriz, a staff member, also reflected on the negative aspects of the participatory process, but linked these with the clarity that the participatory research process had brought:

Beatriz: Negativo? As vezes as coisas não ficam tão claras, trazem mais bagagem para a gente resolver, na verdade é ponto positivo mas é mais trabalhoso. De negativo de fato é ver a posição de algumas pessoas aqui dentro...

Victoria: Como?

Beatriz: De barreiras que pode encontrar, que não deveriam ser assim e hoje estão mais esclarecidas, essas opiniões.

Victoria: E negativo porque agora as pessoas podem ver, e antes estava escondido?

Beatriz: Negativo de saber que tem isso mas no fim das contas é melhor a gente saber e tentar trabalhar para que isso mude mas é ruim ver que as coisas são assim.

Beatriz: Negative? Sometimes things weren't all that clear, it brought more baggage for us to resolve, in truth this is a positive point but it's harder work. What's really negative is to see the position of some people here...

Victoria: How?

Beatriz: The barriers that you can meet, that shouldn't be that way and today they are clearer, these opinions.

Victoria: It's negative because now people can see and before they were hidden?

Beatriz: Negative to know you have this but at the end of the day it's better we know and try to work to change this but it's rubbish to see that things are this way. (Beatriz, staff member)

What Thiago, Susana, Fabio and Beatriz highlight is that the participatory process illuminated some of the hidden personal barriers to the development of children and young people's participation as well as helping the staff to 'better understand what is real' in order to try to change. Crucially, they demonstrate that transparency, both within ourselves and within staff teams, is central to the development of children and young people's participation and highlight that, for them, the value of the research process was that it increased this transparency through exposing underlying questions of both emotion and power.

Exploring the role of the adult: the role of power

The recognition of power is essential to participatory practice (Scott-Villiers 2004). However, despite numerous attempts to theorise power 'among those who have reflected on the matter, there is no agreement about how to define it' (Lukes 2005, p. 61). Valuable contributions, particularly in relation to children and young people's participation, have built on Arendt's conceptualisation of 'power over' developed in her analysis of violence (Arendt 1970) to include four categories: power over, power to, power within and power with (Just Associates 2006). These are particularly useful for understanding power in relation to participation as they provide a framework by which 'people can better understand how forces of subordination and inequity operate in their own lives and envision alternative strategies and visions of power through which they can challenge injustice' (Just Associates 2006, p. 4). Crucially, this framework recognises both the positive and negative dimensions of power through explicitly framing power as relational, dynamic and multidimensional. It is this conceptualisation of power that will frame the understanding of power within this article.

In this research Paulo, the NGO director, identified power as a key issue affecting the practice of participation within the organisation, but he associated this with responsibility:

Paulo: I don't think that most people are prepared for the level of power that they have. It's very worrying for most people.

Victoria: The level of power is in itself worrying for them...?

Paulo: I think it would worry most people, because it, if you accept that you are that powerful it means that you have that much responsibility. So I think that a lot of people, often subconsciously, before even thinking about it [dismiss it] and therefore don't have the responsibility.[1] (Paulo, Director)

As Lukes explains: 'The connection between power and responsibility is "essentially negative: you can deny all responsibility by demonstrating lack of power" (Morriss 2002 cited in Lukes 2005, p. 66) and the "powerful are those whom we judge or can hold to be responsible for significant outcomes"' (Lukes 2005, p. 66). However, this research illustrates that this is a complex process as participants' judgements of who may be responsible for a particular action or outcome did not always correspond. While I may feel powerless to bring about a certain outcome, this does not mean that others will share my view. Essentially, *perceptions of power* are not consistent; whilst I may feel powerless, others may view me as powerful and, consequently, hold me responsible for bringing about a change. Interestingly, this research process clearly indicated that while it may be possible to deny that you have any power to create change, it does not mean that others will not hold you responsible for achieving it.

Within the literature on children and young people's participation, one focus of debate has been the shift from the zero-sum assumption of power that to increase the power of children and young people is to take power away from adults (Hill *et al.* 2004), towards the realisation that participation might be a way to create 'mutually beneficial outcomes in which both increase their power' (Prout and Tisdall 2006, p. 245). However, this research indicated that the challenge is not just about overcoming assumptions about power, but also overcoming *perceptions* of power and consequently perceptions of responsibility. Although issues of power inequalities as a result of organisational hierarchies had been identified as key barriers to the implementation of children and young people's participation, there was a tendency amongst the staff team to look towards the 'other' and place responsibility for change outside of themselves. Power was perceived to lie with others and there was a failure to recognise their own role and relationships of power with those around them. Using the poststructural conceptualisation of power as being socially constructed, and therefore analysis needing to focus upon power relations rather than power itself (Foucault, 2002), I facilitated a team workshop with the objective of exploring this 'passivity' and addressing the 'paranoid fallacy' of assuming that powerlessness results from domination (Lukes 2005, p. 68).

The method used within the staff workshop was adapted from the Venn or Chapati diagramming methodology (see Archer and Cottingham 1996). Developed specifically to address issues of power raised by members of the staff team during interviews, the exercise used circles to represent levels of influence within the organisation. The workshop lasted for 2 h and was attended by 29 staff members, including young people working as youth mediators, administrative staff, educators and members of the management committee. The team was divided according to these four roles to form four groups. My decision to separate the team in this way was deliberate in order to explore the visions and perspectives held by those in different roles within the organisation. Each group was given a large piece of paper with a large circle pre-drawn in the centre. I explained that this circle represented the NGO. I then asked the groups to think about the person or group of people who have the most voice or level of influence within decision-making processes within the organisation and to cut out a circle from the pieces of card that had been provided to represent the size of the voice of this person or group relative to the size of the organisation as a whole (as represented by the pre-drawn circle). The larger the influence the larger the circle. The groups were asked to think about all the different people and groups within the organisation and to place circles within the pre-drawn circle to represent their relative influence in decision-making processes. They were then asked to draw lines between the circles to represent the flows of influence. The groups then repeated the exercise thinking about people or groups having influence within the project's decision-making processes but who are from outside of the organisation, for example, funding bodies or statutory committees. These circles were then placed outside of the pre-drawn circle. At the end of the activity, each group placed their completed diagrams on the wall and discussed and reflected on the various representations of influence.

What emerged most clearly as a result of this activity were the different perceptions of power amongst the team. Within this NGO is a team of educators who work with children and young people classified as being in need of special protection. This multidisciplinary team of educators works intensively with the children and young people and their families to identify and address the young person's needs and accompany them in the process of improving their own lives. What

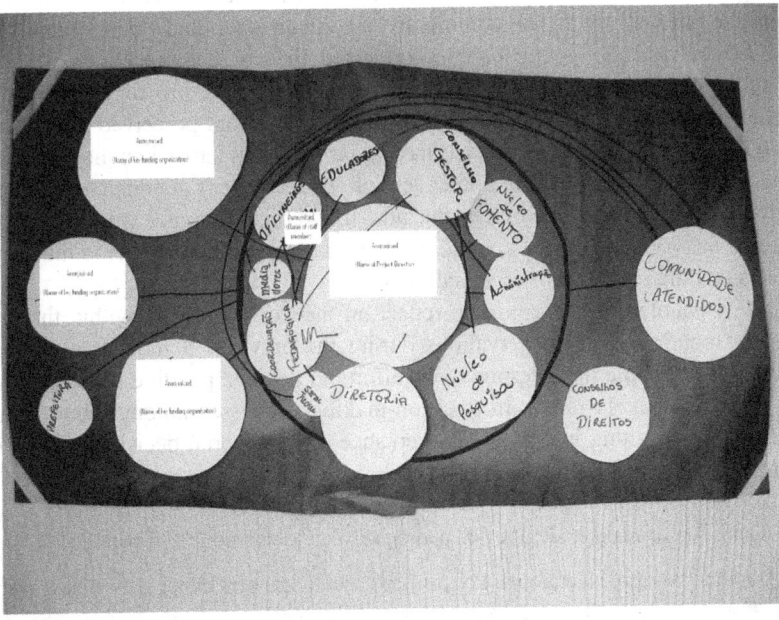

Figure 2. Educators' diagram analysing levels of voice and influence.

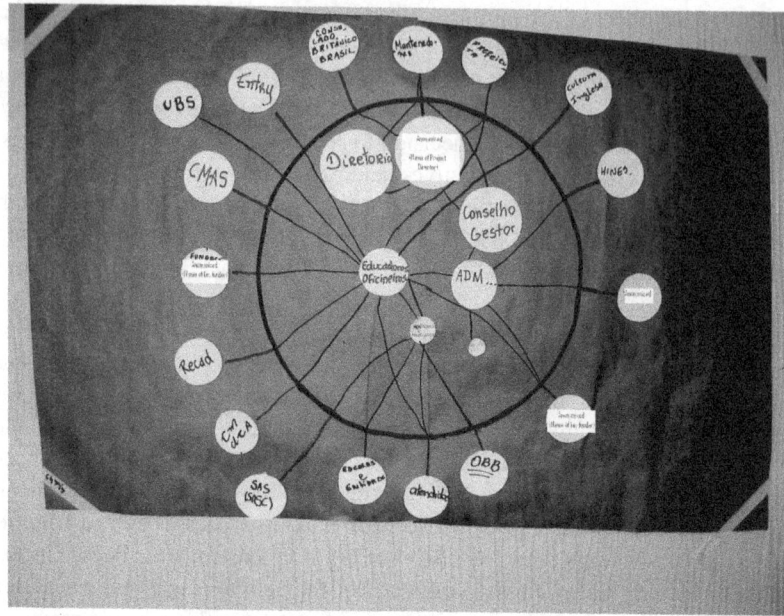

Figure 3. Youth mediators' diagram analysing levels of voice and influence.

was fascinating and also very revealing for both the educators and me was the view that the educators had of themselves (Figure 2) in terms of influence within the organisation compared to the view of the youth mediators (Figure 3). Although the size of the circles representing the educators did not differ significantly within the two diagrams, the flows of influence were significantly different. The educators made only two connections to represent the flow of influence – internally within the organisation they made only one connection, with the coordination team via their team coordinator and externally with the community via the children and young people with whom they work. In contrast, the young people connected the circle representing the educators to well over half of the people and groups of people external to the project as well as connecting them to other people and groups internally. While the educators perceived their own influence as limited, the youth mediators saw them as having far more influence than the educators believed that they had. This was reinforced in a later session with the team of educators to 'interview the diagram' (Kesby et al. 2003). One of the educators found this difference worrying as it confirmed her feeling that the young people with whom she works viewed her as having much more power to make decisions than she feels she actually has.

This activity enabled the educators to reflect on their own position within the organisation. Their reactions confirmed my observations during the previous 3 months working alongside the staff and young people that, despite overt efforts to achieve positive participatory practices, there was a sense of exclusion from decision-making processes amongst both young people and staff members, resulting in a passive acceptance of managerial decisions. Beatriz illustrates the problem:

Victoria: Quais são os maiores desafios da participação na [nome do ONG] para você?

Beatriz: Desafio? Desafio é hierarquia. Porque isso ... as vezes para expor suas idéias, acho que é o maior desafio.

CHILDREN AND YOUNG PEOPLE AS KNOWLEDGE PRODUCERS

Victoria: E como esse desafio podia ser superado?

Beatriz: Superado? Eu acho que persistir na sua idéia, desde que seja coerente, né, unir pensamentos com as outras pessoas, persistir até que fique claro caso não seja aceito ou tenha alguma coisa que vá contra isso, que você persista até que fique claro, e você entenda como coerente isso não acontecer. Não deixa passar, não ficar engolindo sapo à-toa. Já que tem espaço para falar, para articular com outras pessoas, então vamos fazer isso, no fim todos nós temos o mesmo objetivo de atender a comunidade e tem empecilhos que aparecem que são desnecessários e a gente acaba desistindo porque é uma pessoa que está a mais tempo falando sobre isso, tem mais poder de decisão que você e então você acaba desistindo mas acho que não vale a pena desistir assim tão fácil.

Victoria: What for you are the biggest challenges of participation in [name of NGO]?

Beatriz: Challenges? A challenge is hierarchy. Because this... sometimes to express your ideas, I think this is the biggest challenge.

Victoria: And how could this challenge be overcome?

Beatriz: Overcome? I think to persist in your idea, in order to be consistent unify thinking with other people, persist until it's clear, if it's not accepted or has something against it, that you persist until it's clear, and you understand the reason why it won't happen. Don't let it go, don't keep swallowing the same line. Already there is space to speak, to talk with other people, so let's do it, in the end all of us have the same objective to work with the community and there are obstacles that appear that are unnecessary and we end up giving up because someone that has more time [at the organisation] is talking about this, who has more power to make decisions than you and therefore you end up giving up but I think that it's not worth giving up so easily. (Beatriz, staff member)

Beatriz indicates her awareness of a tendency within the staff team to give up because someone may have more experience or power to make the decision. Even though the spaces for debate and discussion existed, staff members tended not to use them. I observed numerous expressions of frustration and anger among colleagues at a similar 'level' in the hierarchy over decisions that had been taken or issues that arose during my time working at this research site. But neither the young people nor the staff felt able to express their frustrations to others and did not involve themselves in the search for a solution; in other words, they did not include themselves as active participants in the process. Returning to the metaphor of the ocean I argue that this 'self' exclusion from the search for solutions creates a disjuncture between the emotional and intellectual currents and the ocean surface. This creates a cyclical effect, shown in Figure 4, whereby the discomfort or insecurity that the person feels at the ocean floor (due to feelings of

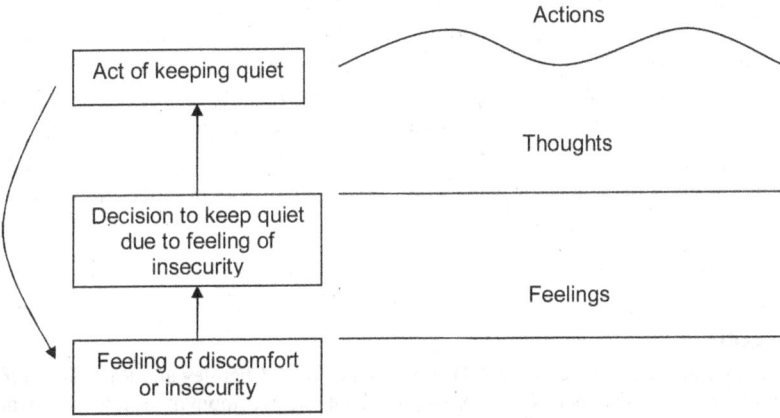

Figure 4. The disjuncture in the dialogical relationship between what we feel, think and do.

exclusion from the decision-making process through feelings of insecurity, for example) is blocked by the intellectual current through the decision to not express their view. However, the decision to not express the view increases the feeling of discomfort through feelings of frustration. Consequently the disjuncture between the two currents increases and the ocean surface becomes more unsettled as a result. The disjuncture could be minimised through the active search for an alternative solution, allowing the intellectual current to acknowledge the emotional, producing more transparent and consistent actions. It is through recognising the dialogical relationship between what we feel, what we think and what we do, and increasing coherence between the elements, I argue, that increases the ability of individuals to include themselves as the active participants of participatory processes.

Conclusion

Active involvement must be founded on a willingness and ability to undertake 'constant vigilance over ourselves' (Freire 2001, p. 51), or what Paula referred to as being able and willing to both face up to our own limits, accepting our own power and consequently accepting our own responsibility. As has been highlighted within feminist geography, reflexivity is a deeply challenging process that requires recognition that 'we are made through our research as much as we make our own knowledge, and that this process is complex, uncertain and incomplete' (Rose 1997, p. 316). Within this article, I have explored the idea of the reciprocal nature of knowledge through the use of Spinoza's ideas of mind and body being of one substance and his ensuing arguments about the increasing or decreasing of potential through active and passive action. Through the metaphor of an ocean I have explored the idea that participatory processes require us to understand our own personal 'oceans' in order to understand how our emotional and intellectual currents affect the transparency of our actions. Crucially, I have argued that each of the elements of our 'oceans' are in dialogical relationship with one another and that, as a result, we need to recognise how each part of the ocean relates and impacts on the others to understand the conditions of the ocean as a whole. Through this I have attempted to further develop the understanding from previous research that 'the process of interaction itself within a participatory process played a key role in changing constructions and exercise of power' (Taylor and Boser 2006, p. 116) and explored the idea that if what we think is inseparable from what we feel, then it is this which affects our ability to relate to people around us – and, critically, also affects other people's ability to relate to us. In short, this article highlights the potential of participatory research to promote reflection on participatory practices, enabling participants to identify and begin to address personal inconsistencies that reduce transparency in personal and professional actions. In essence, I have begun to explore how the participatory process can help to make the invisible barriers to participation visible. Whilst participation is far from being a perfect science and there are no laws to follow, the happy irony of participation is that the process itself can not only provide a transparency to the key issues of power and emotion, a way to make the invisible visible, but it can also be a way to overcome the barrier of incoherence between what we say and what we do.

Acknowledgements

I would like to extend my gratitude to my PhD supervisors, Helen Charnley and Rachel Pain from Durham University, for their ongoing support for this research. I thank the two anonymous reviewers whose thoughtful and informative comments helped further develop my thinking. My sincere thanks to the Economic and Social Research Council for funding this research.

Note

1. Paulo is a native English speaker and therefore this interview was conducted in English.

References

Ahmed, S., 2004. Collective feelings or, the impressions left by others. *Theory, culture and society*, 21 (2), 25–42.
Archer, D. and Cottingham, S., 1996. *The reflect mother manual: a new approach to adult literacy*. London: ActionAid.
Arendt, H., 1970. *On violence*. Orlando, FL: Harcourt.
Askins, K., 2009. 'That's just what I do': placing emotion in academic activism. *Emotion, space and society*, 2 (1), 4–13.
Brown, S. and Stenner, P., 2001. Being affected: Spinoza and the psychology of emotion. *International journal of group tensions*, 30 (1), 81–105.
Cahill, C., 2007. The personal is political: developing new subjectivities through participatory action research. *Gender, place and culture*, 14 (3), 267–292.
Chiu, L.F., 2006. Critical reflection: more than nuts and bolts. *Action research*, 4 (2), 183–203.
Cooke, B. and Kothari, U., 2001. *Participation: the new tyranny?* London: Zed Books Ltd.
Curti, G.H., et al., 2011. For not limiting emotional and affectual geographies: a collective critique of Steve pile's 'emotions and affect in recent human geography'. *Transactions of the institute of British geographers*, 36 (4), 590–594.
Dawney, L., 2011. The motor of being: a response to Steve pile's 'emotions and affect in recent human geography'. *Transactions of the institute of British geographers*, 36 (4), 599–602.
England, K.V.L., 1994. Getting personal: reflexivity, positionality, and feminist research. *Professional geographer*, 46 (1), 80–89.
Foucault, M., 2002. Power. *In*: J.D. Faubion, ed. *Essential works of Michel Foucault 1954–1984, Volume 3*. London: Penguin Books.
Freire, P., 1996. *Pedagogy of the oppressed*. London: Penguin Books Ltd.
Freire, P., 2001. *Pedagogy of freedom: ethics, democracy, and civic courage*. Lanham, Maryland: Rowman and Littlefield.
Glaser, B.G., 1965. The constant comparative method of qualitative analysis. *Social problems*, 12 (4), 436–445.
Goleman, D., 2001. Emotional intelligence: issues in paradigm building. *In*: C. Cherniss and D. Goleman, eds. *The emotionally intelligent workplace*. San Francisco, CA: Jossey-Bass, 13–26.
Gray, B., 2008. Putting emotion and reflexivity to work in researching migration. *Sociology*, 42 (5), 935–952.
Hill, M., Davis, J., Prout, A., and Tisdall, K., 2004. Moving the participation agenda forward. *Children and society*, 18 (2), 77–96.
Jupp, E., 2008. The feeling of participation: everyday spaces and urban change. *Geoforum*, 39 (1), 331–343.
Just Associates, 2006. *Making change happen: power. Concepts for revisioning power for justice, equality and peace*. Washington, DC: Just Associates.
Kapoor, I., 2005. Participatory development, complicity and desire. *Third world quarterly*, 26 (8), 1203–1220.
Kesby, M., 2005. Retheorizing empowerment-through-participation as performance in space: beyond tyranny to transformation. *Signs: journal of women in culture and society*, 30 (4), 2037–2065.
Kesby, M., Kindon, S., and Pain, R., 2003. 'Participatory' approaches and diagramming techniques. *In*: R. Flowerdew and D. Martin, eds. *Methods in human geography: a guide for students doing a research project*. London: Longman, 144–166.
Kraftl, P. and Horton, J., 2007. 'The health event': everyday, affective politics of participation. *Geoforum*, 38 (5), 1012–1027.
Laurie, N., et al., 1999. Working with genders and geographies. *In*: N. Laurie, C. Dwyer, S. Holloway and F. Smith, eds. *Geographies of new femininities*. Essex: Pearson Education Ltd, 41–66.
Lukes, S., 2005. *Power: a radical view*. Basingstoke: Palgrave Macmillan.
Lupton, D., 1998. *The emotional self: a sociocultural explanation*. London: Sage Publications Ltd.
Mannion, G., 2010. After participation: the socio-spatial performance of intergenerational becoming. *In*: B. Percy-Smith and N. Thomas, eds. *A handbook of children and young people's participation: perspectives from theory and practice*. Oxon: Routledge, 330–342.
Moreau, P.F., 1996. Spinoza's reception and influence. *In*: D. Garret, ed. *The Cambridge companion to Spinoza*. New York: Cambridge University Press, 408–434.

Moser, S., 2008. Personality: a new positionality? *Area*, 40 (3), 383–392.

Nicholls, R., 2009. Research and indigenous participation: critical reflexive methods. *International journal of social research methodology*, 12 (2), 117–126.

Pain, R., 2004. Social geography: participatory research. *Progress in human geography*, 28 (5), 652–663.

Percy-Smith, B., 2007. 'You think you know? ... you have no idea': youth participation in health policy development. *Health education research*, 22 (6), 879–894.

Percy-Smith, B. and Thomas, N., 2010. Conclusion: emerging themes and new directions. *In*: B. Percy-Smith and N. Thomas, eds. *A handbook of children and young people's participation: perspectives from theory and practice*. Oxon: Routledge, 356–366.

Pile, S., 2010. Emotions and affect in recent human geography. *Transactions of the institute of British geography*, 35 (1), 5–20.

Pile, S., 2011. For a geographical understanding of affect and emotions. *Transactions of the institute of British geographers*, 36 (4), 603–606.

Pinkey, S., 2011. Participation and emotions: troubling encounters between children and social welfare professionals. *Children and society*, 25 (1), 37–46.

Prout, A. and James, A., 1997. *Constructing and reconstructing childhood: contemporary issues in the sociological study of childhood*. 2nd ed. London: Falmer Press.

Prout, A. and Tisdall, E.K.M., 2006. Conclusion: social inclusion, the welfare state and understanding children's participation. *In*: E.K.M. Tisdall, J.M. Davis, M. Hill and A. Prout, eds. *Children, young people and social exclusion: participation for what?* Bristol: The Policy Press, 235–246.

Punch, S., 2011. Hidden struggles of fieldwork: exploring the role and use of field diaries. *Emotion, space and society*, doi: 10.1016/j.emospa.2010.09.005 (Online publication complete 4th November 2010).

Reason, P. and Torbert, W.R., 2001. Toward a transformational science: a further look at the scientific merits of action research. *Concepts and transformations*, 6 (1), 1–37.

Rose, G., 1997. Situating knowledges: positionality, reflexivities and other tactics. *Progress in human geography*, 21 (3), 305–320.

Sawaia, B., 2009. Psicologia e desigualdade social: uma reflexão sobre liberdade e transformação social. Unpublished paper. Pontifica Universidade Católica, São Paulo.

Scott-Villiers, P., 2004. Personal change and responsible well-being. *In*: L. Groves and R. Hinton, eds. *Inclusive aid: changing power and relationships in international development*. London: Earthscan, 199–209.

Taylor, P. and Boser, S., 2006. Power and transformation in higher education institutions: challenges for change. *IDS bulletin*, 37 (6), 111–121.

Thien, D., 2005. After or beyond feeling? A consideration of affect and emotion in human geography. *Area*, 37 (4), 450–456.

Uprichard, E., 2008. Children as 'being and becomings': children, childhood and temporality. *Children and society*, 22 (4), 303–313.

Widdowfield, R., 2000. The place of emotions in academic research. *Area*, 32 (2), 199–208.

Wyness, M., 2009. Adults' involvement in children's participation: juggling children's places and spaces. *Children and society*, 23 (6), 395–406.

Taking the long view: temporal considerations in the ethics of children's research activity and knowledge production

Kate Hampshire[a], Gina Porter[a], Samuel Owusu[b], Simon Mariwah[c], Albert Abane[c], Elsbeth Robson[a], Alister Munthali[d], Mac Mashiri[e], Goodhope Maponya[f] and Michael Bourdillon[g]

[a]*Department of Anthropology, Durham University, Durham, UK;* [b]*Department of Population and Health, University of Cape Coast, Cape Coast, Ghana;* [c]*Department of Geography & Regional Planning, University of Cape Coast, Cape Coast, Ghana;* [d]*Centre for Social Research, Chancellor College, University of Malawi, Zomba, Malawi;* [e]*Independent Transport Consultant, Pretoria, South Africa;* [f]*CSIR, Pretoria, South Africa;* [g]*Department of Sociology, University of Zimbabwe, Harare, Zimbabwe*

> Children are increasingly engaged in the research process as generators of knowledge, but little is known about the impacts on children's lives, especially in the longer term. As part of a study on children's mobility in Ghana, Malawi and South Africa, 70 child researchers received training to conduct peer research in their own communities. Evaluations at the time of the project suggested largely positive impacts on the child researchers: increased confidence, acquisition of useful skills and expanded social networks; however, in some cases, these were tempered with concerns about the effect on schoolwork. In the follow-up interviews 2 years later, several young Ghanaian researchers reported tangible benefits from the research activity for academic work and seeking employment, while negative impacts were largely forgotten. This study highlights the unforeseeable consequences of research participation on children's lives as they unfold in unpredictable ways and underscores the temporal nature of children's engagement in research.

Introduction

It is now commonly accepted (in theory at least) by researchers and other organisations that children are not merely passive recipients of adult knowledge, but actively create, interpret and produce meanings, understandings and 'knowledge'. Children's right to have their views heard and taken seriously is enshrined in the UN Convention on the Rights on the Child (UNCRC) (OHCHR 1989). Specifically, Article 12 states that

> the child who is capable of forming his or her own views has the right to express those views freely in all matters affecting the child, the views of the child being given due weight in accordance with the age and maturity of the child.

Article 13 states that 'The child shall have the right to freedom of expression; this right shall include freedom to seek, receive and impart information and ideas of all kinds...'.

However, the practice is often less straightforward, and many attempts to elicit and integrate 'children's voices' and children's knowledge have been criticised for tokenism (see James 2007).

Christensen and Prout (2002) identified four contrasting perspectives that underpin current research with children, reflecting Hart's (1992) well-known 'ladder of participation'. Traditionally, most social science research has cast children as 'objects': acted on by others, vulnerable, in need of adult protection and insufficiently competent to participate in research. Knowledge about children is, therefore, generated largely from parents, teachers and other adult caretakers. The second perspective acknowledges children as persons with subjectivity, but attempts to use children as 'informants' are strongly conditioned by judgements about their cognitive abilities and social competence. The third (relatively new) strand extends the recognition of children's subjectivities to see children as social actors in their own right: effective research respondents, no less *inherently* competent than adults.

Finally, there is a small but growing interest in children becoming *researchers in their own right*. Mirroring earlier feminist arguments about research for and by women (e.g. Oakley 1981), it has been argued that knowledge about children's worlds is best produced *by children*. Advocates of this approach claim that it leads to better quality research since 'children are the primary source of knowledge about their views and experiences' (Alderson 2001, p. 151). In our child mobility study (see below), we noted that child researchers were better able than adults to elicit children's emotional and corporal experiences of movement and travel and to uncover instances of physical abuse (Robson et al. 2009, Porter et al. 2010). But research by children is not merely about the quality of research outcomes; it is widely portrayed as part of a wider 'political struggle for recognition, representation and equality' (Jones 2004, p. 114), helping to redress adult–child power imbalances in the research process (Kellett et al. 2004), and a way of 'rescu[ing] [children] from silence and exclusion, and from being represented, by default, as passive objects' (Alderson 2001, p. 142).

However, bringing children into research generates a number of practical, epistemological and ethical questions, many of which relate to more general critiques of participatory research (Christensen 2004, Hampshire et al. 2005, Gallagher 2008). First, much child-centred research has been criticised as being tokenistic, failing to really challenge generational hierarchies (Kellett et al. 2004, James 2007, Schäfer and Yarwood 2008). Power differentials, which shape adult–child interactions and thus research outcomes, may be downplayed or even ignored (Christensen 2004, James 2007). Such research, based on crude and superficial notions of 'empowerment', risks manipulation and promotes a form of 'ethnographic ventriloquism' (Geertz 1988, p. 145, cited by James 2007, p. 263), in which adults' views are expressed 'through' children and thereby receive an artificial sense of authenticity. A second, related, point is the risk of representing 'children's voices' as an authentic, unproblematic 'truth'. As we have noted, 'it is a naïve and dangerous form of essentialism to attribute uncritically particular rapport, or connection, to the biological attribute of age' (Robson et al. 2009, p. 475, see also James 2007).

Third, there can be a serious risk of privileging the views and perspectives of certain children over those of others. Although it is widely noted that children do not constitute a homogenous group with a common body of 'knowledge' and a common set of experiences, needs and desires (e.g. James 2007), the conceptualisation of 'children's knowledge' (like 'indigenous knowledge'; Sillitoe 1998) risks glossing over the diversity of children's lives and experiences. Indeed, the UNCRC itself, which sets out the rights of 'the child', risks doing the same. Differences between children, based on age, gender, class, wealth, disability, family situation and personality (among other things), are bound up in complex relations of power (Christensen and Prout 2002, Schäfer and Yarwood 2008, Crivello et al. 2009), which shape the nature of knowledge that might be produced from 'child-led' research, highlighting particular voices and representations while obscuring others.

Finally, there are serious ethical questions about children's research engagement; it is these that we focus on here. How does being a child researcher impact on children's rights? Is it always in children's best interests to participate in research? Roberts (2000, p. 238, cited by James 2007, p. 268) observed that

> the reasons why a child or a young person should choose to participate in research are clearer in some studies than others ... we cannot take it for granted that participation in research and the development of increasingly sophisticated methods to facilitate children's participation is necessarily in their best interests. (See also Schäfer and Yarwood 2008, p. 124)

As noted by Pyer (2008), research carried out with the best of intentions can have unintended consequences. In our child mobility study, we had lengthy debates (with children and adults) about whether the participation of child researchers might infringe their rights to education and protection from exploitation (UNCRC Articles 28 and 32; see Robson et al. 2009, Porter et al. 2010); we expand on these concerns below.

Almost all evaluations of the impact on children of participation in, or conducting, research have been carried out during or immediately after the research activities (although see Johnson 2010 for a notable exception). Numerous potential benefits have been identified, including increased self-confidence, research/communication skills and self-efficacy (Alderson 2001, Kellett et al. 2004, Porter and Abane 2008, Schäfer and Yarwood 2008, Robson et al. 2009, Johnson 2010, Porter et al. 2010), alongside some potential costs: social, economic and others (Porter and Abane 2008, Robson et al. 2009, Porter et al. 2010). However, little is known about what happens *after* the project has finished and the adult researchers have gone away. Are any of the supposed benefits to children sustained or do they diminish over time? Might short-term benefits, such as increased confidence and aspirations, turn into long-term difficulties, as aspirations are not met and new-found status diminishes? And what about the costs – do they translate into long-term disadvantages or not?

This paper offers a small contribution to addressing these questions. We draw on our child mobility study conducted in Ghana, Malawi and South Africa (www.durham.ac.uk/child.mobility), which involved child-to-child, and adult-to-child, research. During the 3-year project, we evaluated the process and its impacts on the child researchers (Robson et al. 2009, Porter et al. 2010). Two years later, a sub-sample of children in Ghana were re-interviewed to find out what had happened to them since and what (if any) had been the impact of the project on their subsequent lives.

The child mobility study and child researchers

The aim of the child mobility study was to understand mobility constraints faced by girls and boys and how these might impact on livelihood opportunities and wellbeing. This involved exploring children's lived experiences of moving around their neighbourhoods, on daily journeys to school, to markets, to fetch water and firewood, to farms, to friends' homes and to other places where they needed or wanted to go. The research was conducted in three phases. Phase 1, which we focus on here, entailed recruiting and training child researchers[1] in each country to conduct research into mobility among their peers in their own communities. These findings were used to inform phase 2, in which adult academic researchers conducted qualitative research in eight field sites per country. In phase 3, a questionnaire survey was administered (by adults) to 3000 children to test the hypotheses arising from phases 1 and 2.

We have described elsewhere in detail how the child researcher phase of the project was conducted (Robson et al. 2009, Porter et al. 2010); the key points are summarised here. Adult academic collaborators in each country recruited child researchers through local schools in urban, peri-urban and rural settlements. In some cases, the selection process involved pupils being asked to write an essay on transport/mobility. Consent to participate was sought from the children themselves and from parents/guardians and school heads. All the child researchers were in-school children and were identified (by their teachers or via the essay) as being academically able enough to undertake the training and research.

Although the overall project was designed by adults, we tried to ensure that the child researchers planned their own part. Nineteen children (from all three countries) participated in the inception workshop, where research objectives and methods were refined and ethical guidelines drawn up. We adopted a participatory approach to ethics (Pain 2008): child and adult researchers working together developed a 12-point Code of Practice, which stated that 'Children should benefit from being researchers' (Robson et al. 2009). This was followed by six training workshops, each about a week long (two per country), facilitated by adult academics, in which children were introduced to the project in more detail and were taught a range of relevant research methods. In groups, the children then decided on their specific research topics (within the overall transport/mobility theme), methods and time frame for conducting the research. The most popular methods were activity and travel diaries, photo-journals, accompanied walks and individual interviews, with some children also choosing to run focus group discussions, and ranking and counting exercises.

Altogether, 70 young researchers (YRs) (most aged 10–18 years) conducted fieldwork over periods ranging from 3 weeks to 2 months. They worked mostly in pairs, mainly within their own settlements and neighbourhoods. In-country adult researchers provided support via regular (usually weekly) visits and phone calls. The child researchers were also taught some simple data analysis techniques and were encouraged to discuss and reflect on their findings. They received small financial payments for their work and were allowed to keep various fieldwork materials: wristwatches (for timekeeping), folders, notebooks and copies of photographs.

At the final project workshop, 19 YRs (12 Ghanaians, 4 Malawians and 3 South Africans) produced a short book, drawing together their findings and those of their colleagues (available online at http://www.dur.ac.uk/child.mobility/children_mobility_book_webversion.pdf). Four thousand copies have been printed and presented (by child and adult researchers) in Ghana and Malawi (funded by the Africa Community Access Programme) to governmental and non-governmental stakeholders, including the child researchers' schools and communities. All the 70 YRs were awarded certificates of workshop attendance.

Impacts of being researchers on the children: immediate findings

During the project, the UK-based lead researcher interviewed every child researcher available to elicit his or her views and experiences associated with the research. Forty-one child researchers were interviewed across the three countries: 12 in Malawi, 11 in Ghana and 18 in South Africa; the findings from these interviews have been published: Porter et al. (2010). Another publication described the experiences of the Malawian child researchers during the project: Robson et al. (2009). Here, we summarise those findings to provide a point of comparison for the follow-up work presented below.

Positive experiences

During the project, all the child researchers interviewed were generally very positive about their experiences. They all claimed to find the work interesting, enjoyable and informative. Prominent in their accounts was increased self-confidence, particularly in communicating with others. As one 14-year-old rural South African girl put it, 'The project has shaped me to be able to talk to people nicely' (Porter et al. 2010, p. 219), while 17-year-old Samuel[2] from Ghana commented that 'It helped me about talking in public, meeting new people'. This increased self-confidence was evident to the adult researchers: at the inception workshop, the children were generally quietly spoken and many were afraid to voice their opinions; by the final workshop, all the child researchers present contributed enthusiastically and vocally to discussions, even when being filmed for Ghana national television.

Child researchers also commented positively on their interactions with the adult academics and research assistants (RAs), who were co-opted as friends and confidantes. Several made phone calls, office visits and occasional emails to particular RAs (Porter et al. 2010, p. 219). This opportunity to build personal networks with (possibly powerful and influential) adults outside their home communities was an important part of the project for some and was influential in widening horizons and opening a new sense of possibility, especially for the older children: 'It's the first time I've had interactions with professors and lecturers and the university campus and it made us aware we too can come here' (Kofi, 17 years, Ghana). Several children thought that they had learned skills from the adult researchers which might prove useful in the future. For example, one 14-year-old South African girl commented that 'The research taught me how to talk to people and now we can research on other things' (Porter et al. 2010, p. 219), while 16-year-old John from Ghana remarked that 'I liked working with adults. It makes you more learned'.

Finally, payment was important for the child researchers. This issue generated substantial debate among the adult research team, balancing the need for fair remuneration (to avoid exploitation), with labour legislation requirements and consistency between countries (Robson et al. 2009, Porter et al. 2010). While the sums paid were not large, children clearly valued these highly and put them to good use. One 16-year-old Malawian boy, for example, told us that 'the money which I was given has helped me to buy text books for school' (Robson et al. 2009, p. 473), while another 15-year-old Malawian boy reported using his earnings to buy fertiliser to contribute to household food production.

Negative experiences

While all the child researchers evaluated their experiences positively, some also pointed to difficulties. Most prominent were the time/opportunity costs of attending workshops and conducting research. The logistical difficulties of coordinating a three-country study meant that some workshops and research activities took place during school term time for some children. Although the timetables were agreed upon in advance with the children, parents and teachers, some secondary-school children, in particular, worried about the impact on their schoolwork. As one 15-year-old Malawian girl put it, 'What I did not like about doing the research is when we are doing our meetings or research while our friends are learning and I do not want to miss my classes' (Robson et al. 2009, p. 471). Indeed, one Malawian child was not granted parental permission to participate because of concerns about missing schoolwork. Some secondary-school pupils in Ghana, particularly girls, expressed similar concerns: 'it was difficult combining with school work. I'd prefer the holidays' (Florence, 16 years); 'it was a problem about school work a bit; it would be better in the holidays or weekends' (Rebecca, 16 years).

For some, research work also limited the time available for domestic work or income-generating activities. However, most parents valued and supported their children's participation in the project, and other household members sometimes helped with household chores to free up the child researcher's time, as Florence (16-year-old, Ghana) explained: 'My parents were very happy about it – working with the University of Cape Coast. My sister took over household chores from me'. To what extent this impacted negatively on the workloads and wellbeing of siblings is unclear, although this should be recognised as a possibility. For others, balancing time for research with other demands was more difficult: 'Yes [there was a problem with a clash with house work] because when I wanted to do it [the project], they [parents] would say I must do household first... But I managed it and finished everything in time' (14-year-old South African girl; Porter et al. 2010, p. 223). One Ghanaian girl had to withdraw from the project because it was incompatible with her selling cassava, which provided vital household income (this was the only case of a YR withdrawing from the project).

Child researchers, like adults, can face difficulties and sometimes unpleasantness: abrupt refusals from potential respondents, occasional rudeness, demands for money and even the threat of physical violence. During training workshops, we worked through various such scenarios. Very few children reported problems dealing with respondents, but there were some exceptions. One South African girl said that 'They [adults in the village] were saying this is nonsense and even swearing at us' (Porter et al. 2010, p. 221), while a 16-year-old Ghanaian boy commented that '[Some of the adults] asked questions... "why is it important?" Sometimes they accuse you – "it's no need to ask us questions like that"' (Porter et al. 2010, p. 222). Demands for payment for photos and interviews were occasionally made to children in all three countries, which they understandably found difficult to handle.

Finally, there are the more mundane, but nevertheless real, day-to-day fieldwork hazards, such as illness, adverse weather conditions, transport difficulties and other practicalities. Long accompanied walks in the rainy season, which involved getting wet and muddy, were mentioned by some, as well as fears of ferocious dogs or snakes encountered during fieldwork:

> the worst was the accompanied walk because you don't know where they are going and when you've finished you have to accompany them to wherever else they are going. (Kofi, 17 years, Ghana)

Some others encountered unforeseen research expenses, which they had to meet out of pocket, such as transport costs or – in the case of one Malawian boy – soap to get clean after a long muddy walk home from fieldwork (Robson et al. 2009, p. 473).

Long-term evaluations and experiences of the child researchers

The evaluations discussed above reflect the experiences of child researchers during or shortly after their fieldwork. This has led us (and others) to speculate optimistically about the potential future benefits to children participating as researchers, 'which we believe are immense and far reaching' (Robson et al. 2009, p. 478; see also Schäfer and Yarwood 2008, Porter et al. 2010). How far are these speculations justified?

Since the project ended, we have maintained contact with many of the YRs in all the three countries, and several have actively kept in touch with us (as we encouraged them to do at the final project meetings). In Ghana, follow-up interviews were conducted with as many YRs as could be contacted in October–December 2009 (2–3 years after completion of fieldwork and 12–14 months after the final workshop). Eight of the original 16 were interviewed face to face; two more were followed up by telephone. To minimise potential bias, the interviewer was an adult unknown to the young people. Interviews were semi-structured and covered various themes relating to experiences of the project (training workshop, conducting the research and analysis, attending the final workshop and writing the book) and their subsequent personal and professional lives. Summary information about the eight interviewees is given in Table 1.

All eight YRs were keen to be re-interviewed and spoke enthusiastically about the project. All described positive impacts on their subsequent lives, ranging from self-confidence to new career or educational opportunities. The final workshop was a particular highlight for several:

Table 1. Summary information for Ghana ex-YRs in September–December 2009.

Name	Current age	Age at the start of the project	Sex	Location	Current activities
Rebecca	19 years	15 years	F	Cape Coast	Completed SHS[a], working in a shop while attending computer literacy classes; hopes to read law at university
Mercy	19 years	15 years	F	Cape Coast	Completed SHS, not currently working
Kwesi	20 years	16 years	M	Cape Coast	Completed SHS, currently attending Cape Coast Technical Institute; hopes to become a military officer
Samuel	20 years	16 years	M	Sunyani	Completed SHS, studying for BSc in Agricultural Engineering at KNUST (Kwame Nkrumah University of Science and Technology, Kumasi)
Kofi	20 years	16 years	M	Sunyani	Completed SHS, studying for BSc in Natural Resource Management at KNUST
Florence	19 years	15 years	F	Sunyani	Completed SHS, currently teaching children (3–4 years) in kindergarten; applied for tertiary education – nursing qualification
Abena	19 years	15 years	F	Sunyani	Completed SHS and pre-university diploma in IT; plans to do a 3-year journalism degree
Elizabeth	16 years	12 years	F	Sunyani	Currently in second year at SHS; wants to study nursing after completion

[a]SHS, senior high school.

> It was very exciting, I liked everything! The building, the people, everything! The workshop was good – everyone's idea was accepted. The visit to the castle was very good – everything! ... [Re training workshop] I loved it – I wanted to stay even longer! (Florence)

Self-confidence, pride and status

One of the strongest themes to emerge from all the follow-up interviews was an increased sense of self-confidence from conducting research and participating in workshops. Several commented that this had been of enduring benefit, enabling them to communicate effectively in new situations:

> At first I was a very shy person, but because of skills in interviewing, I can approach people and talk to them freely. ... This project ... has taken my shyness away. Now I don't feel unnecessarily shy as I used to be and I think this will stay with me forever. The project has made me more confident in whatever I do. The confidence alone can open opportunities for me ... If it were those days [before the project], I could not even look straight into your face during the interview, but now, even though it is my first time of meeting you, I am comfortable to speak freely to you. (Mercy)

> The skills have given me the confidence to talk to people. Initially when I wanted to talk to an adult, I became a bit tense up but now, as you can see, I can talk freely without shyness or tension. ... The confidence and patience that I acquired throughout the project are going to be with me throughout my life. ... This makes it really easy for me to fit in any place I go, no matter who I meet. (Kwesi)

Contributing to this self-confidence was an enormous pride in their achievements, particularly producing the book:

> I have written a book! I didn't know it would be so easy to write a book! I am the author of a book! (Florence)

> This book will be seen by people all over the world and it has my name on it. When anyone buys it they will see my name! (Abena)

This sense of pride came out far more strongly in the follow-up interviews than in the earlier ones, possibly because the book had not yet come out, so the results felt less tangible. All the interviewees had kept copies of their research work, and three brought the copies, along with their certificates, to the interviews:

> I still go through all my notes – it makes me proud. I've brought all my notes from the project to university with me, I've told all my friends here about the book. (Samuel)

For some, participation in the project had led to elevated status within their communities:

> Now people [from home town] respect me a lot. The UCC [University of Cape Coast, the local collaborating institution] bus came to pick us up. They [neighbours] thought I'm a big person! Now people in town come to ask me for advice. (Kofi)

Personal contacts and networks

Another very prominent theme in the follow-up interviews was the new friends made: other YRs in Ghana, academics from UCC[3] and children from Malawi and South Africa:

> The workshop in Mankessim was a wonderful experience because I made a lot of friends from different parts of Ghana and Africa. For example, I met a friend from Malawi – Ali. He taught me some words in his native language and I also taught him some words in my language. ... The next day I drew a cartoon of Ali and I showed it to him ... He became very excited and we became friends from that moment. (Kwesi)

> [Biggest impact of the project is] how we related to each other. People from Malawi, South Africa, we all came together. We are all at one level ... It was the first time I ever saw a South African face. Before that I thought they were all whites. (Florence)

Many of these contacts had been sustained. All eight interviewees reported being still in touch with other Ghanaian YRs, as the accounts given below illustrate:

> After the project, I used to contact [X] and she called me frequently. Nowadays we do not contact each other so frequently. ... But these friendships are important to me and even now I know that any time I go to Sunyani[4] I have a friend there. (Mercy)

> I am still in contact with Samuel and Abena, and I see them sometimes. (Florence)

> I still see Cyril – he is here at KNUST[5], and I've been calling Mercy. (Samuel)

Almost all were still in touch with adult RAs, who continued to give them advice about their studies and other matters. These relationships were highly valued, both affectively and instrumentally, for providing sound advice and as a possible route to future university admission:

> We had a good relationship with the adult researchers because they were very friendly. After the project, [X] used to call me frequently to ask me about my school and the progress in my studies. [Y] also used to call me to advise me to study hard. In fact, all of them became our friends. (Mercy)

> I visited [X] at UCC – we went out for lunch. And I keep in touch with [Y]. I phone him. He gives me advice – how to study well. (Samuel)

Fewer had stayed in touch with YRs from overseas, but Samuel said that he occasionally phoned one of the South African YRs after the workshop, and Abena had kept up a written correspondence with a Malawian YR. Lack of phone credit was reported as an obstacle to maintaining international friendships, as was keeping correct contact information:

> After the workshop, I could not contact him because the telephone number he gave me could not be reached. But he also gave me his email address and so I will try it one of these days. (Kwesi)

Interestingly, none of the YRs reported any negative impacts (e.g. jealousies) of their research participation on relationships with friends, siblings or others, even when asked a direct question about this.

Academic and career-related impacts

The biggest downside of the project identified by YRs at the time of the research was the time costs and associated impact on their schoolwork. When asked specifically about negative aspects, three re-voiced these concerns:

> The only negative thing is that I had to miss lessons in class in order to participate in the project. Sometimes I found it very difficult to catch up with my colleagues when it comes to the lessons I missed, and I would be too tired to study in the evenings when I came back from fieldwork. (Kwesi)

> [If doing the project again] we would stagger the work – we need to be given more time. We were at school, so the only time we had to work on the project was after school. (Kofi)

> The only problem was time. Because we did the research on schooldays, and we did the interviews after school, and got harassed and embarrassed at school for not doing our homework. (Abena)

However, with a couple of years' hindsight, all the YRs thought that their research experience would help them in their future studies, and some were already using their skills for university work:

> Now the skills [learned in the project] help me to seek information from other people. In the future, I want to study law, so I think the research skills will help me to collect information from people. (Rebecca)

> At first, my friends thought I would be wasting my time [doing this project]. They said instead of learning, I would be moving from place to place, which could affect my studies negatively. After the project, my friends are now happy that I took part ... they realised my involvement in the project was a good thing. (Mercy)

> [The research skills] help with field trips that we do with our university courses. ... It also helped me find out about the admissions process here [KNUST]. (Kofi)

All the YRs also believed that their knowledge, experiences, skills and having the certificate would help them in their future careers (see also Schäfer and Yarwood 2008):

> The project has taught me to be a good listener and also to become more patient with people. I want to be a military officer and I believe that being a good listener will help me handle people effectively. (Kwesi)

> I have had my certificate laminated. I am going to be a journalist. I have learned so many things how to interview people, how to get their story. Because when a journalist comes it is after the incident has happened, so you need to know how to interview the person, how to ask questions. ... I can do these

things. I will show them [prospective employers] my certificate and they will see, oh, she has been to a workshop, she knows how to do these things.... The certificate, it will help me. If I have that and another person doesn't, the employers will choose me. (Abena)

When you show them this certificate, they cannot say you are just small-fry! ... That certificate will help me in nursing. In nursing you have to talk to people and do research. (Elizabeth)

Moreover, four YRs said that they had already used their experience and/or certificates to help gain employment or other opportunities:

I used the certificate when I got a [temporary] job as a school-teacher. I showed it to the headmistress in the interview. (Abena)

I went to work in a hospital before I started university. I showed them my certificate and I put it on my CV. (Kofi)

I have become the Welfare Officer in my department [at KNUST] because I told them about the research. I try to solve the other students' problems. (Samuel)

At the interview for the teaching job [currently nursery teacher] I told the headmistress about the research. (Florence)

Discussion

The follow-up interviews suggest that the YRs' evaluations of their research experiences were more unequivocally positive a couple of years down the line than at the time. Problems identified during fieldwork (missing school, lack of time for homework or domestic chores, problematic respondents and practical difficulties) had been largely forgotten or were no longer considered substantial issues. The three interviewees who mentioned clashes with schoolwork did so only in response to a direct question; it was not mentioned spontaneously by any. And, with hindsight, none thought that there had been any lasting damage: the issue was more the short-term pressure of catching up with schoolwork when they were tired or the embarrassment of incomplete homework than the long-term negative impacts on academic performance.

We should perhaps not be surprised at the YRs' successes since completing the project. Although they lived in resource-poor settlements, they were all in-school children selected, at least in part, based on their academic abilities. They may well have gained university admission or formal-sector employment without participating in the project. Nonetheless, it is noteworthy that their concerns about the negative aspects of participating in the project were relatively short-lived.

We are not suggesting that long-term evaluations of research participation necessarily matter more than immediate ones. The stress of having to catch up with school work, fears of ferocious dogs or unpleasant refusals from prospective participants are very real and can be extremely distressing, even if they do not carry lasting detrimental consequences. However, our work does imply that the full impacts (positive, negative or neutral) of engaging children in research cannot always be known at the time and that there may be important things to learn from following up over a longer period.

If we continue to follow up the YRs over coming years, the picture may change again. As noted above, they were extremely proud of their certificates and had very high hopes of their instrumental value. While three interviewees had apparently used their certificates to gain employment, it remains to be seen whether these will indeed be useful in their quest to obtain permanent jobs as journalists, military officers or nurses. If these expectations and aspirations are not met, will that lead to later disillusionment? The perhaps unrealistic nature of some aspirations was evident in a couple of off-the-cuff remarks: Kofi, only half-joking, commented that with his certificate, he would be able to get admission to UCC, while Abena's mother (whom we contacted in order to find Abena) asked (again, half-joking) why her daughter could not gain admission to a UK university with her certificate.

In an earlier publication, we argued that Article 3 of the UNCRC, which states that 'In all actions concerning children ... the best interests of the child shall be a primary consideration', is highly problematic in its interpretation (Robson et al. 2009). First, the identification of 'best interests' involves resolving often conflicting values among both adults and children about the importance of, for example, formal education versus informal learning experiences. Second, there is a difference between 'best' in an ideal world and 'making the best' of a messier set of economic and social realities in which children find themselves. We have argued that '"best" has to be seen in relative rather than absolute terms – the best of what is realistically possible or available under specific economic, technological and social conditions' (Robson et al. 2009, p. 470–1). A similar point can be made in relation to 'protection from exploitation', which is rarely grounded in the children's lived experiences and may overlook the role of wider structural inequalities in shaping children's work (Robson et al. 2009; see also Bey 2003, Nieuwenhuys 2005). In other words, we and others have argued that interpretation of children's rights needs be situational rather than absolute, making the best of the messy and often non-ideal realities of children's lives, which may involve complex negotiations of competing and priorities and interests. This is not an excuse for ethical sloppiness; rather it is an attempt to work towards what Christensen and Prout (2002, p. 492) described as 'a practical, value-oriented ethics' (see also Horton 2008, Pain 2008).

Here, we take this argument a stage further, suggesting that the interpretation of children's rights must also adopt a *temporal* perspective. Although the UNCRC acknowledges that rights and responsibilities shift with 'the evolving capacities of the child' (Article 5), the impression given is one of linear and predictable change over time. In practice, 'best interests' and 'exploitation' are dynamic concepts, and perspectives on what constitute either of these may shift in unpredictable ways over the course of, and beyond, childhood. And because the shifts are not predictable, it is difficult to weigh up the relative importance of immediate and possible long-term impacts. If a child feels upset after an unpleasant encounter with a potential respondent, or misses out on earning extra household income as a result of undertaking academic research, to what extent should these experiences be offset against the future, necessarily uncertain, possibility that increased confidence and skills acquired will enhance future opportunities and well-being? With the benefit of hindsight, it became easier for the YRs in our project to identify which impacts were lasting or fleeting and which made a material difference to their lives.

The main focus of this paper has been to apply a temporal lens to the ethics of children's research participation. However, it is also worth reflecting on the extent to which our work has addressed other criticisms levelled at some children's research initiatives, highlighted in the introduction. These are part of a 'wider ethics' that relates not just to individual researchers' experiences, but also to how children's views are represented and communicated to policy-makers in the context of inter- and intra-generational power relations and concerns about social justice.

Most obviously, perhaps, selecting child researchers based (partly) on their academic abilities has inevitably meant privileging the views and perspectives of these children at the expense of others. The rationale for this was largely practical: the time and resource constraints of the project meant that children had 1 week (the training workshop) to acquire the necessary skills to conduct semi-independent research, which entailed grasping some quite complex concepts quickly. This highlights another issue about children's research: the extent to which it must fit within an 'adult' research paradigm. In order to be valued by the academic community and policy-makers, the children's research had to fit a particular model, using well-established methods (interviews, focus groups and various other techniques that now comprise the participatory research canon). We thought that the process of socialisation into this research paradigm would be easier and quicker for children who had already demonstrated their ability to 'achieve' in adult-valued ways through academic success. This observation raises another series of questions, which we are not able to address here, about whether young people's research can offer the opportunity to rethink research paradigms: whether their contribution can only be valued if they conduct research as 'proto-adults' or if new approaches to doing research and establishing validity and authority might be possible.

Our research design, which combined child- and adult-led research, helped us to address the related criticism of children's 'voices' being accepted uncritically as authentic representations of 'truth'. At the planning stage, we agonised about whether it was necessary to incorporate adult-led research: what would adults be able to find out about children's lives that children could not? Our decision to use adult researchers was based partly on practical considerations: the children's time constraints, combined with logistical and safety issues entailed in them conducting all the research in 24 field sites, precluded this possibility, due to the size and scope of the project. However, there were also epistemological considerations at stake. Were the views and experiences that the child researchers uncovered any more 'real' or 'true' than those identified by the adult researchers? Our analysis of the resulting research materials leads us to suggest that each can offer different *kinds* of insights. As noted above, child researchers elicited much more detailed accounts of children's emotional and corporal experiences of movement and travel and uncovered some instances of physical abuse (Robson et al. 2009, Porter et al. 2010). However, the adult researchers were able to draw out a wider range of perspectives from both children and adults on, for example, mobility constraints in different settlement types. Indeed, it was often at the intersection between the insights generated by the child and adult researchers that more nuanced understandings of the complex, negotiated and embodied nature of children's mobilities emerged.

The design of the child mobility study also helped us to address some accusations of tokenism often apparent in child-led research. Because the adult research component was premised on the findings from the children's research, the adult researchers had to engage actively and critically with these. Of course, inter-generational power differentials were not eliminated in this process; however, we believe that doing the child-led research first, so that it could shape the rest of the project rather than being an addendum, was important in redressing them.

Conclusion

This is a very small-scale study, based on follow-up interviews with 10 YRs in Ghana. It is possible that the six Ghanaian YRs who were not contactable might have had different views; for one thing, none was still in touch with the adult researchers. However, neighbours and acquaintances said that they had gone away to study or work, so there is no reason to suppose that their experiences were significantly worse than those we interviewed. The subsequent experiences of the Malawian or South African YRs might also have been different from those of their Ghanaian counterparts, although *ad hoc* news from them suggests that they too are, by and large, doing well.[6] In Malawi, for example, several of the YRs are now in higher education or formal-sector employment. Three that we know of are at university in Malawi. A fourth gained a university scholarship in Sudan, and another moved to South Africa with her husband, both of them obtained their passports through the project. However, some are just 'sitting' (not in employment or education), while one has just had to suspend his tertiary education for lack of fees.

It is not our intention, therefore, to draw general conclusions about the directions in which young people's experiences of research participation might change over time. Children's lives unfold in nonlinear and unpredictable ways, which make it impossible to foresee the full future consequences, positive, negative or neutral, of being part of a research project. Instead, we invite other researchers to take the temporalities in the ethics of children's knowledge production seriously.

What does this mean in practice for adult researchers trying to work in collaborative ways with children? First, we must expect that things will change, during and after research projects in unknowable ways: in the words of Agnew (2006, p. 4), we must be open to 'being ... surprised by what the world throws up', and be ready to respond to the ethical challenges that this generates. We should not expect that ethical guidelines drawn up at the start of a project will continue to apply unproblematically; instead, we should be prepared to re-visit and re-negotiate these as the often messy realities of people's lives unfold. As Bauman (1993, p. 12, cited in Horton 2008, p. 368) observed: 'human reality is messy and ambiguous – and so moral decisions, unlike abstract ethical principles, are ambivalent. It is in this sort of world that we must live'. Even over the course of the project, as the YRs faced important examinations, changes in family situations and other unforeseen events, we had to continually rethink, on an individual basis, how to achieve the aspiration, enshrined in the project Code of Practice, that 'children should benefit from being researchers'.

Second, our work underscores the point that participatory research is not a one-off event, whose effects are frozen at a single point in time. If participation achieves its goal of being an empowering, and therefore potentially life-changing, experience, we should not expect that changes will all happen overnight. Any of us who gets involved in participatory research with children must, therefore, recognise that we are in it for the long haul. We cannot simply do the research, leave and hope that things will turn out OK. Moreover, the rapid expansion in the availability of mobile phones and other communication technologies sets up a facility for, and often an expectation of, sustaining research-generated relationships far beyond the end of a project. As Coleman and Collins (2006, p. 5) have noted, 'in a world of inter-connections, we never leave the field'. While we may not always be in a position to continue to provide emotional and other kinds of support to ex-child researchers once a project has finished, by sticking around, at least on the end of a phone or email, we can at least continue to observe how things turn out, insights that may help us to become more sensitive to the dynamic, temporal and 'evental' (Horton 2008, p. 365) nature of children's knowledge production in future work.

Notes

1. We acknowledge that the term 'child' is problematic and culturally contingent (Dehne and Riedner 2001). However, at the opening workshops, the 'child researchers' agreed that this was the term they wanted to use. Two years later, at the final workshop, their consensus had shifted towards 'young researchers'. Here, we use these terms interchangeably.
2. Pseudonyms are used throughout.
3. University of Cape Coast, Ghana.
4. Sunyani is a large town, 8 hours' bus journey from Cape Coast, where Mercy lives.
5. Kwame Nkrumah University of Science and Technology, Kumasi.
6. Tragically, one of the South African young researchers has since died.

References

Agnew, J., 2006. Open to surprise? *Progress in human geography*, 30 (1), 1–4.
Alderson, P., 2001. Research by children. *International journal of social research methodology*, 4 (2), 139–153.
Bauman, Z., 1993. *Postmodern ethics*. Oxford: Blackwell.
Bey, M., 2003. The Mexican child: from work with the family to paid employment. *Childhood*, 10 (3), 287–299.

Christensen, P.H., 2004. Children's participation in ethnographic research: issues of power and representation. *Children and society*, 18 (2), 165–176.

Christensen, P. and Prout, A., 2002. Working with ethical symmetry in social research with children. *Childhood*, 9 (4), 477–497.

Coleman, S. and Collins, P., 2006. *Locating the field: space, place and context in anthropology.* Oxford: Berg.

Crivello, G., Canfield, L., and Woodhead, M., 2009. How can children tell us about their wellbeing? Exploring the potential of participatory research approaches within *Young Lives. Social indicators research*, 90 (1), 51–72.

Dehne, K.L. and Riedner, G., 2001. Adolescence – a dynamic concept. *Reproductive health matters*, 9 (17), 11–15.

Gallagher, M., 2008. 'Power is not an evil': rethinking power in participatory methods. *Children's Geographies*, 6 (2), 137–150.

Hampshire, K., Hills, E., and Iqbal, N., 2005. Power relations in participatory research: a case study from Northern England. *Human organization*, 64 (4), 340–349.

Hart, R., 1992. *Children's participation: from tokenism to citizenship.* Florence: UNICEF.

Horton, J., 2008. A 'sense of failure'? Everydayness and research ethics. *Children's geographies*, 6 (4), 363–383.

James, A., 2007. Giving voice to children's voices: practices and problems, pitfalls and potentials. *American anthropologist*, 109 (2), 261–272.

Johnson, V., 2010. Are children's perspectives valued in changing contexts? Revisiting a rights-based evaluation in Nepal. *Journal of international development*, 22 (8), 1076–1089.

Jones, A., 2004. Involving children and young people as researchers. *In*: S. Fraser, V. Lewis, S. Ding, M. Kellett, and C. Robinson, eds. *Doing research with children and young people.* London: Sage, 113–131.

Kellett, M., et al., 2004. 'Just teach us the skills please, we'll do the rest': empowering ten-year-olds as active researchers. *Children and society*, 18 (5), 329–343.

Nieuwenhuys, O., 2005. The wealth of children: reconsidering the child labour debate. *In*: J. Qvortrup, ed. *Studies in modern childhood: society, agency, culture.* Basingstoke, UK: Palgrave Macmillan, 167–183.

Oakley, A., 1981. Interviewing women: a contradiction in terms? *In*: H. Roberts, ed. *Doing feminist research.* London: Routledge, 30–61.

OHCHR (1989). *Convention on the Rights of the Child* [online]. Office of the United Nations High Commissioner for Human Rights. Available from: http://www2.ohchr.org/english/law/crc.htm [Accessed 12 March 2012].

Pain, R., 2008. Ethical possibilities: towards participatory ethics. *Children's geographies*, 6 (1), 104–108.

Porter, G. and Abane, A., 2008. Increasing children's participation in African transport planning: reflections on methodological issues in a child-centred research project. *Children's geographies*, 6 (2), 151–167.

Porter, G., et al., 2010. Children as research collaborators: issues and reflections from a mobility study in sub-Saharan Africa. *American journal of community psychology*, 46 (2), 215–227.

Pyer, M., 2008. Unintended consequences? Exploring the un(fore)seen effects and outcomes of research. *Children's geographies*, 6 (2), 213–217.

Roberts, H., 2000. Listening to children; and hearing them. *In*: P. Christensen and A. James, eds. *Research with children.* London: Falmer Press, 225–241.

Robson, E., et al., 2009. 'Doing it right?': working with young researchers in Malawi to investigate children, mobility and transport. *Children's geographies*, 7 (4), 467–480.

Schäfer, N. and Yarwood, R., 2008. Involving young people as researchers: uncovering multiple power relations among youths. *Children's geographies*, 6 (2), 121–135.

Sillitoe, P., 1998. The development of indigenous knowledge: a new applied anthropology. *Current anthropology*, 39 (2), 223–252.

Index

Note:
Page numbers in **bold** type refer to figures
Page numbers in *italic* type refer to tables
Page numbers followed by 'n' refer to notes

Abane, A.: et al 2–4, 89–102
accommodation: rented 12–13
actors 21–7, 34, 71; altered initial motivations 27; individual motivations 35; social 2, 5–7, 17, 90
adult mediation: children's voices 42, 50, 53
adult-child: dichotomy 71; power imbalance 90
adult-youth/child relationships 2–4, 32, 34, 54n, 59–60, 63, 64
adults: emotions/power impact on participation 71–88
Africa 1–4, 43; Averting New Variant Famine project 43; dissemination workshop 45, **47**, *52*; Lesotho 2, 39–43, **47**, 48–50; Malawi 2–4, 39–51, **47**, 54n, 89–97, 100; southern 39–56; West 49, *see also* sub-Saharan Africa
Agnew, J. 101
Ahmed, S. 73
AIDS (Acquired Immunodeficiency Syndrome) 3; impacts and participatory methods (southern Africa) 39–56
Alderson, P.: and Goodey, C. 62
Allen, J. 66
Ansell, N.: et al 2, 39–56
Arendt, H. 81
Askins, K. 73
asylum seekers 51

Bailey, M.: and Wills, W. 60
Balen, R.: et al 63
Bartky, S. 65
Bauman, Z. 101
Blazek, M.: and Hraňová, P. 1–4, 21–38
Blerk, L. van: et al 2, 39–56
Bourdillon, M.: et al 2–4, 89–102
Boyden, J.: and Ennew, J. 41–3
Bragg, S. 59–62
Bratislava *see* participatory video project (Slovakia)

Brazil 3, 71–4; participatory process and adult emotions/power impact 71–88; São Paulo 3, 71–4
British Federation for Detached Youth Work 36n
Burningham, K.: et al 2–3, 5–20

Cahill, C. 42
Cairns, L.: and Williamson, B. 60
Cameron, J.: and Gibson, K. 42
Cartesian conceptualisations 72
Cartesian understandings 73
child mobility 4, 91–2, 100; study 90–1
child researchers: sub-Saharan Africa 89–102
child services 2, 18; participation (UK) 57–70
children's voices 42, 50, 53, 61, 62, 66–7, 89–90, 93, 100; adult mediation 42, 50, 53
Chimombo, S. 49
Chiu, L.F. 77
Christensen, P.: and Prout, A. 90, 99
chronic illness 43, *see also* AIDS
Clark, A. 57–8, 63–4
clean-up process 14
client-expert relationships 59
Coleman, S.: and Collins, P. 101
collaborative research 22–5
community 23–6, 31–3, 89; development projects 23, 28–31; dissemination workshops 48; resilience 5, 18, *see also* participatory video project
consumerism 61
contamination: floodwater 15, *see also* storyboard methodology
contextual relationships 2
Cook, T.: and Hess, E. 63–4
criticality 57, 61, 67

Darlington 59
data analysis software (Atlas Ti) 9
data production: collective 53; methods 46
Davies, G. 23
Delueze, G.: and Lukes, Z. 72
Derrida, J. 59

INDEX

dialogical pedagogy 71
dialogical relationship **76, 85**
dichotomy 71; adult-child 71
disaster recovery 12
disasters: natural 6, 17
disempowerment 61, 67
disruption: flooding 5–6, 9–12, 16
Durham Council 59–60, 64

Edgetown Secondary School (Hull) *8*
educational psychology service (EPS) 1, 57–9; Darlington 59–60
effects 6
embeddedness 2
emotions 71–88; adult and participatory process 71–88; coherent 75–8; participation 71–88; role 72–4; Spinozian view role 72–4
empowerment 1, 22, 40, 75, 90; dis- 61, 67
England 5–6, 25; Pitt Review (2008) 6
English (language) 51–2
Ennew, J.: and Boyden, J. 41–3; *et al* 54
epistemological perspective 40
epistemology: participatory 40; post-structuralist 42
ethics: child research activity 89–102
ethnographic ventriloquism 1, 90
European Union (EU) 36n; Youth in Action programme 36n
Every Child Matters strategy 17
exclusion: social 66

feminist geography 72, 75–6, 86
Flicker, S.: and Guta, A. 63
flood narratives: experiences 2–3, 5, 12, 14; learning from (UK) 5–20
flood recovery 5–6, 16–17; clean-up process 14; experiences 12; impacts 6; long-term 7–10; process 5–6, 10, 16–18; rented accommodation 12–13; stress 16
flooding 5–7, 16; disruption 5–6, 9–12, 16; effects 6; implications 13; media coverage 9; perspectives 5; physical health effects 6, 10; secondary 2, 13–14; social effects 6; vulnerable sub-group 5
flooding impacts 1–3, 5–6, 17; emotional 6; hidden 14; social 5–7, 16
floodwater 9, 15; contamination 15
food security/insecurity 43
Foucault, M. 3, 59
Freire, P. 75

Gallagher, M. 34, 42
geography: feminist 72, 75–6, 86
Ghana: child researchers 4, 89, 91–4, *95*, 96, 100; University of Cape Coast (UCC) 93–6, 99
Gibson, K.: and Cameron, J. 42

Goodey, C.: and Alderson, P. 62
Gray, B. 77
Gunewardena, N. 17
Gunter, H.: and Thomson, P. 64–6
Guta, A.: and Flicker, S. 63

Hajdu, F.: *et al* 2, 39–56
Hampshire, K.: *et al* 2–4, 89–102
Hart, R. 41, 61, 90
hazards: natural 5
health: physical 6, 10
Hess, E.: and Cook, T. 63–4
heterogeneity 2, 61
HIV (Human Immunodeficiency Virus) 43
Ho, E. 54n
Horton, J.: and Kraftl, P. 75
Hraňová, P.: and Blazek, M. 1–4; 21–38
Hull (UK): child flood narratives 5–20; City Council 8
humanistic epistemologies 42

illness: chronic 43 *see also* AIDS
inclusion: social 66
individual resilience 5
inequalities: power 17, 41, 54n, 62, 75, 82, 99
international youth exchange (Liverpool) 4, 25, 30–4
Investing in Children (IiC) 58–60, 64–5; Durham Council 57, 59

James, A. 1
Jupp, E. 75
Jupp Kina, V. 3, 71–88

Kambewa, D.: *et al* 48
Kapoor, I. 76
Kesby, M. 42
Kindon, S. 21–3
Kingston-upon-Hull (UK): flood narratives 5–20
Kohli, R.K.S. 51
Kraftl, P.: and Horton, J. 75

Langevang, T. 54n
Lesotho 2, 39–43, **47**, 48–50
Liverpool: international youth exchange 4, 25, 30–4
loss 14
Lukes, S. 82
Lukes, Z.: and Delueze, G. 72
Lunch, N.: and Lunch, C. 22
Lupton, D. 73

McDowell, L. 24
Malawi 2–4, 39–51, **47**, 54n, 89–97, 100
Manchester University 1
Mannion, G. 75

INDEX

Maponya, G.: *et al* 2–4, 89–102
Mariwah, S.: *et al* 2–4, 89–102
Marshside Primary School (Hull) 8
Mashiri, M.: *et al* 2–4, 89–102
Mayer, V. 35
Medd, W.: *et al* 2–3, 5–20
media coverage: flooding 9
mobility: child study 89–91, 92, 100
Moore, R.: and Muller, J. 66
Moran-Ellis, J.: *et al* 2–3, 5–20
Moser, S. 77
motivations 27, 35
Muller, J.: and Moore, R. 66
Munthali, A.: *et al* 2–4, 89–102
Murdoch, J.: and Pratt, A.C. 62

narratives: flood 5–20
natural disasters 6, 17
natural hazards 5
neo-liberalism 62–3
networks: social 51, 89
Nicholls, R. 77
non-governmental organisations (NGOs) 1–3, 74, 80–3; child-focused 1; São Paulo 3
non-profit organisation (NPO): Ulita 25

Office for Standards in Education (Ofsted) 62
orphanhood 49
orphans 40–3, 48–9
Owusu, S.: *et al* 2–4, 89–102

Parr, H. 21–3, 36n
participation 1; children's services 57–70; emotions 71–88; research 40, 89
participatory action research (PAR) framework 21–4
participatory activities 40–5, 51
participatory design 63–4, 67
participatory epistemology 40
participatory methods 3, 39–56, 63; AIDS impacts (southern Africa) 39–56; collective data production 53
participatory practice 3, 74, 81, 86; positive 84
participatory process: and adult emotions/power impact (Brazil) 71–88
participatory research 40–5, 52–3, 59, 72, 80, 86, 101; epistemological critiques 41–2; justifications 40–1
participatory video project (Slovakia) 2–4, 21–38; Bratislava 2, 21, 25, **25**; collaborative research dynamics 22–5; community screening **30**; field project **33**; intersubjective diversity 21; intersubjective process 26–33; motivations 22, 32–5; power relations 24; video-making process 21–35; video-training **28**

pedagogy: dialogical 72
perspectives 1; adult-focused 6; disciplinary/interdisciplinary 1; epistemological 40; flooding 5
Peters, P.E.: *et al* 48
physical health: flooding effects 6, 10
Pile, S. 73
Porter, G.: *et al* 2–4, 89–102
post-modernism 63
postructuralist epistemologies 42
poststructuralism 42
poverty 8, 16, 48, 63
power 2, 39–42; adult and participatory process 71–88; adult-child imbalance 90; conceptualisation 82; dynamics 29–30; inequalities 17, 41, 54n, 62, 75, 82, 99; modern 59; positive/negative dimensions 82; pre-existing dynamics 2; relations 2, 24, 35, 41–4, 52, 90, 99; role 81–6
Pratt, A.C.: and Murdoch, J. 62
pre-school settings 58, 64
Prinsloo, J.: and Tomaselli, K.G. 24
Prout, A.: and Christensen, P. 90, 99
psychological resilience 7, 18
Punch, S. 75
Pyer, M. 91

Reason, P.: and Tortbert, W.R. 77
Reay, D. 64
reflexivity 65, 71–88
relationships 2, 2–4, 21–38, 39; client-expert 59; complex 22; contextual 2; emerging 21–38; inter/intra-generational 4; researcher-participant 22; social 2, 39; youth/child-adult 2–4, 32, 34, 54n, 59–60, 63, 64
rented accommodation 12–13
research 1–2; assistants (RAs) 93–6; collaborative dynamics 22–5; design 74; local assistants 3, 50; long-term evaluations/experiences 94–8; objects 2; participation 40, 89, *see also* participatory research
research activity: ethics 89–102
research gap 57–70
researcher-participant relationships 22
researchers: child 89–102
resilience 4, 5–9, 13–17; community 5, 18; forms 5, 14–16; individual 5; psychological 7, 18
responsibilisation 61
Roberts, H. 91
Robertson, C.: and Shaw, J. 22
Robson, E.: *et al* 2–4, 39–56, 89–102
Royal Geographical Society/Institute of British Geographers (RGS/IBG) 1–2

INDEX

São Paulo 3
Save the Children 1; briefing paper (2000) 1
Schäfer, N.: and Yarwood, R. 61
Scott-Villiers, P. 75
secondary flooding 2, 13–14
services: child 2, 18, 57–70
seventeenth century 72
Shaw, J.: and Robertson, C. 22
Slovakia *see* participatory video production
social actors 2, 5–7, 90; individual 17
social effects: of flooding 6
social exclusion 66
social inclusion 66
social networks 89; maps 51
social relationships 2, 39, 66
social-work practitioners 1
South Africa 4, 89–97, 100
Spinoza, B. 3, 71–80; concept 77; philosophy 72–3
Spinozian view 72–4; role of emotions 72–4
storyboard methodology 5–11, 15, 48–51; emotional 50–1; flood journey 9, **9, 11, 13, 15**
stress 16
sub-Saharan Africa 1; child researchers 89–102; participatory process and impacts of AIDS 39–56

Tapsell, S.: *et al* 2–3, 5–20
temporality 3–4, 12–13, 17; and ethics of child research 89–102
Thomson, P.: and Gunter, H. 64–6
Todd, L. 2, 57–70
tokenism 2, 41, 61, 66, 89–90, 100; counter 61
Tomaselli, K.G.: and Prinsloo, J. 24
Torbert, W.R.: and Reason, P. 77

United Kingdom (UK) 1–3, 6, 18, 92; child services 2, 18, 60; children's services participation in evaluation 57–70; Darlington 59; Durham Council 59–60, 64; Edgetown Secondary School *8*; Every Child Matters strategy 17; Hull 2, 5–7, 12; Hull City Council 8; Investing in Children (Durham) 57; Investing in Children (IiC) 58–60, 64–5; learning from flood narratives 5–20; Liverpool 4, 25–7, 30; Manchester University 1; Marshside Primary School *8*
United Nations (UN) 1, 5, 17; Convention on the Rights of the Child (1989) (CRC) 1, 17, 89–90, 99
University of Cape Coast (UCC) 93–6, 99

Valentine, G. 6
ventriloquism: ethnographic 1, 90
video 21, 26–34; making process 21–35; product 21, 27, 34; tool 23, 27; training **28**, *see also* participatory video project
voices 83, **83, 84**; children's 42, 50, 53, 61, 62, 66–7, 89–90, 93, 100
vulnerability 2–4, 6–10; demographic criteria 14; flooding 5; hidden 12–14; new forms 2–4, 14–16; pre-existing 17; socio-economic criteria 14
vulnerable sub-group 5
Vygotsky, L. 72–3

Wales 6
Walker, M.: *et al* 2–3, 5–20
Walker, P.: *et al* 48
Warshak, R. 40
Western rationality 54n
Western tradition 54n
Whittle, R.: *et al* 2–3, 5–20
Whitty, G.: and Wisby, E. 61
Williamson, B.: and Cairns, L. 60
Wills, W.: and Bailey, M. 60
Wisby, E.: and Whitty, G. 61

Yarwood, R.: and Schäfer, N. 61
youth exchange: international 4, 25, 30–4
youth/child-adult relationships 2–4, 32, 34, 54n, 59–60, 63, 64

www.routledge.com/9780415834377

Related titles from Routledge

Diverse Spaces of Childhood and Youth: Gender and socio-cultural differences
Edited by Ruth Evans and Louise Holt

Diverse Spaces of Childhood and Youth focuses on the diverse spaces and discourses of children and youth globally. The chapters explore the influence of gender, age and other socio-cultural differences, such as race, ethnicity and migration trajectories, on the everyday lives of children and youth in a range of international contexts. These include the diverse urban environments of Istanbul, Copenhagen, Helsinki, Toronto, London, and Bratislava and the contrasting rural settings of Ghana and England. The analyses of children's, young people's, parents' and professionals' experiences and discourses provide critical insights into how gender and other socio-cultural differences intersect. Overall, the book provides an original contribution to geographies of children, youth and families and research on diversity and difference in global contexts.

This book was published as a special issue of *Children's Geographies*.

August 2013: 246 x 174: 240pp
Hb: 978-0-415-83437-7
£85/$145

For more information and to order a copy visit
www.routledge.com/9780415843377

Available from all good bookshops

www.routledge.com/9780415567565

Related titles from Routledge

Young People, Class and Place

Edited by Robert MacDonald, Tracy Shildrick and Shane Blackman

Under the weight of apparently growing consumer affluence, globalisation and post-modern social theory, many have proclaimed the declining significance of social class and place to young people's lives – and for social science. Drawing upon new, empirically grounded, theoretically innovative studies, this volume begs to differ.

It argues that the youth phase provides a privileged vantage point from which to interrogate and think about broader processes of social change and social continuity. The book's chapters consider the problems of growing up in gang neighbourhoods and young people's use of space for leisure; new patterns of class formation and youth transition in Eastern Europe; the effects of classed labels and identities (such as 'chav' and charver') in youth culture and schooling; the changing meanings of class and place for young women in changing socio-economic landscapes; new patterns of youth culture and transition among Black young men in East London; and how we think and theorise about change and continuity in youth studies.

This book was based on a special issue of *Journal of Youth Studies*.

February 2012: 216 x 138: 152pp
Hb: 978-0-415-56756-5
£85 / $145

For more information and to order a copy visit
www.routledge.com/9780415567565

Available from all good bookshops